INVENTION AND INNOVATION

Previous works by the author

China's Energy
Energy in the Developing World (edited with W. Knowland)
Energy Analysis in Agriculture (with P. Nachman and T. V. Long II)
Biomass Energies
The Bad Earth
Carbon Nitrogen Sulfur
Energy Food Environment
Energy in China's Modernization
General Energetics
China's Environmental Crisis
Global Ecology
Energy in World History
Cycles of Life
Energies
Feeding the World
Enriching the Earth
The Earth's Biosphere
Energy at the Crossroads
China's Past, China's Future
Creating the 20th Century
Transforming the 20th Century
Energy: A Beginner's Guide
Oil: A Beginner's Guide
Energy in Nature and Society
Global Catastrophes and Trends
Why America Is Not a New Rome
Energy Transitions
Energy Myths and Realities
Prime Movers of Globalization
Japan's Dietary Transition and Its Impacts (with K. Kobayashi)
Harvesting the Biosphere
Should We Eat Meat?
Power Density
Natural Gas
Still the Iron Age
Energy Transitions (new edition)
Energy: A Beginner's Guide (new edition)
Energy and Civilization: A History
Oil: A Beginner's Guide (new edition)
Growth
Numbers Don't Lie
Grand Transitions
How the World Really Works

INVENTION AND INNOVATION

A BRIEF HISTORY OF HYPE AND FAILURE

VACLAV SMIL

THE MIT PRESS
CAMBRIDGE, MASSACHUSETTS
LONDON, ENGLAND

This book was set in ITC Stone and Avenir by New Best-set Typesetters Ltd. Printed and bound in the United States of America.

Library of Congress Cataloging-in-Publication Data

Names: Smil, Vaclav, author.
Title: Invention and innovation : a brief history of hype and failure / Vaclav Smil.
Description: Cambridge, Massachusetts : The MIT Press, 2023. | Includes bibliographical references and index.
Identifiers: LCCN 2022019899 (print) | LCCN 2022019900 (ebook) | ISBN 9780262048057 (hardcover) | ISBN 9780262374255 (epub) | ISBN 9780262374262 (pdf)
Subjects: LCSH: Inventions—Defects—History—Popular works. | System failures (Engineering) —History. | New products—History. | Errors—History. | Technological forecasting.
Classification: LCC T15 .S63 2023 (print) | LCC T15 (ebook) | DDC 600—dc23/eng/20220909
LC record available at https://lccn.loc.gov/2022019899
LC ebook record available at https://lccn.loc.gov/2022019900

10 9 8 7 6 5 4 3 2 1

CONTENTS

1

INVENTIONS AND INNOVATIONS

A LONG HISTORY AND MODERN INFATUATION

The evolution of our species is a history of physical and behavioral changes closely tied to the outcomes of invention. Invention is a big umbrella that covers items belonging to four principal categories. The first category comprises an enormous variety of simple handmade items, starting with stone tools made once our ancestors became bipedal, which freed their hands for performing deliberately complicated tasks. The progress of their toolmaking was, as best as we can measure it from excavations or discoveries in caves, very slow. The oldest crude stone tools came more than three million years ago, larger, well-crafted (bifacial) hand axes and cleavers followed only about 1.5 million years ago, wooden stone-tipped spears appear to be about half a million years old, and only about 25,000 years ago did the Upper Paleolithic hunters master the artisanal production of an array of composite tools, including adzes, axes, harpoons, needles, and saws, and accompanying pottery.

The widespread adoption of crop cultivation was predicated on the invention of numerous farm tools. The domestication of horses for riding started with bits and bridles (stirrups and saddles came much later). Draft animals required many specific designs for their harnessing to plows, carts, or wagons—collars, reins, traces, bellybands for horses, yokes for oxen. All sedentary societies engaged in, and some excelled at, making wooden furniture, designing and firing pottery, and smelting ores to produce tools and weapons. Modern societies still depend on a profusion of such simple products, including hammers and saws, wooden chairs and benches, and cups and plates, but only a tiny share of their production is now artisanal as machines have taken over.

Machines belong to the second category of inventions, that of new and more or less complex devices or mechanisms deployed for both stationary use and transportation. Large waterwheels, windmills, tall stone blast furnaces with waterwheel-activated leather bellows, and oceangoing sailing ships were among the most remarkable premodern inventions in this category. By the late nineteenth century, the Sears catalogues listed thousands of such items, ranging from pocket watches to small sewing and large wheat-threshing machines, and recent product arrays offer repeated examples of excess: we now have more than a thousand models of mobile phones on the global market and in the US about seven hundred distinct models of passenger road vehicles (I cannot say cars anymore, as new vehicles are mostly SUVS, pickups, and vans).

New ideas have to be embodied—be it in simple practical tools, in a complex machine, or in the even more complex machine assemblies that make up modern industrial enterprises and are now often highly automated: carmaking factories are perhaps the best common examples of such aggregations, with robots doing everything from carrying and positioning parts to welding and painting. Readily available stones and wood can be turned into only a limited range of tools, machines, and structures. That is why the third category of inventions, new materials, has been an obvious marker of civilization's progress, from the age of stone and wood to the era of metals, mixtures, and compounds. Inventions in the third category began with bronze, proceeded to iron and steel (iron's largely decarbonized alloy), and now include aluminum and a dozen other common metals, as well as glass, cement (an aggregate of materials), and, starting in the late nineteenth century, a still-expanding variety of plastics and—the most recent addition—carbon-based composites, light yet stronger than steel.

The fourth category of invention consists of new methods of production, operation, and management, ranging from marginal but economically rewarding improvements to fundamentally new and highly automated ways of mass-scale manufacturing, information gathering, and data processing. One of the most remarkable and most consequential inventions of this kind was Michael Owens's glass bottle–making machine, introduced in 1904. For centuries, bottles had to be blown individually, and in the late nineteenth century came the first semiautomatic machines: in either case, these operations employed children to carry and handle molten glass and release it from forms. By 1899 more than seven thousand American boys were employed in these hot and dangerous conditions, as captured in contemporary photographs: only child labor in deep coal mines was similarly appalling. In contrast, Owens's machines gathered glass directly from the furnace and the entire process was done without any human labor. Even Owens's earliest model was able to make 2,500 bottles every hour, compared to 200 bottles per hour for semiautomatic setups (fig. 1.1).

After World War II almost every established way of mass-scale industrial production was transformed—made more efficient, cheaper, faster—by the introduction of electronic controls (now embedded in every new rice cooker or coffee maker), and electronics had an even greater impact on data acquisition, processing, and dissemination. During World War II the terms *calculators* and *computers* were used for (mostly younger) women employed in tedious data entry and processing; now every small laptop has data-processing power far superior to the most advanced pre-microprocessor computers of the late 1960s, and the selection of electronic machines ranges from miniature monitoring devices, some small enough to be affixed to the backs of flying insects, to giant data servers, built, owing to their incessant high electricity demand, near an inexpensive electricity supply.

As commonly used, the meanings of the terms *invention* and *innovation* have a large overlap, but innovation is perhaps best understood as the process of introducing, adopting, and mastering new materials, products, processes, and ideas. Accordingly, there could be plenty of invention without commensurate innovation, with the USSR being perhaps the best recent example of the dissonance. Soviet scientists had many notable inventions to their credit, with eight of them receiving Nobel Prizes (including Landau and Kapitsa for low-temperature physics and Basov and Prokhorov for lasers and masers), and the prioritized, heavily financed military R&D efforts made the country's weaponry competitive with US advances.

The USSR amassed 45,000 nuclear warheads. The MiG-29 and Su-25 were among the world's best fighter planes ever deployed in combat, and when American engineers were designing the world's first stealth aircraft,

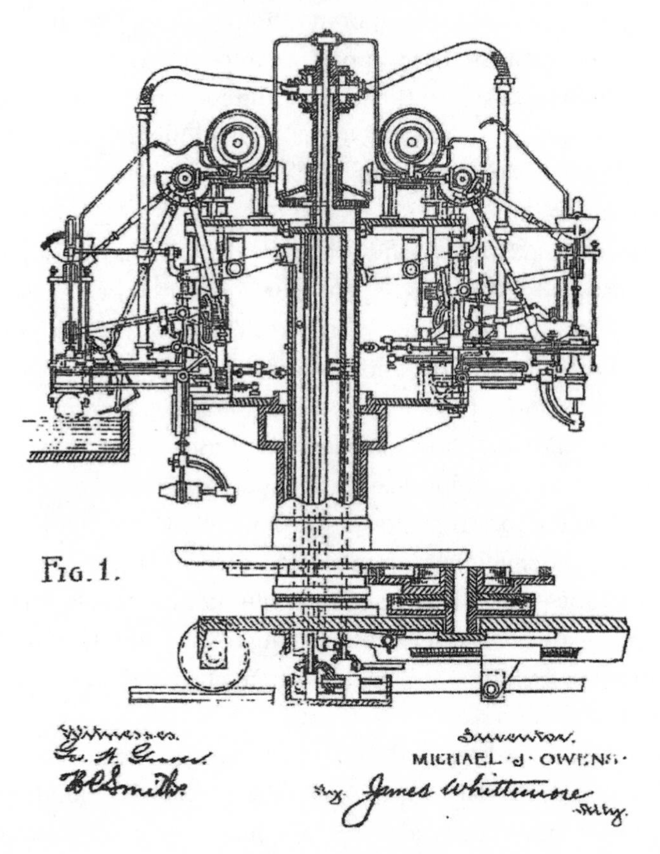

Figure 1.1 Michael Joseph Owens's glass-shaping machine. US patent filed by Toledo Glass Company. *Source:* M. Owens, Glass-shaping machine (US Patent 766,768, filed April 13, 1903, and issued August 2, 1904), https://patents.google.com/patent/US766768.

they used Pyotr Ufimtsev's equations for predicting the reflections of electromagnetic waves from the plane's surfaces. The USSR also excelled in the world's most important energy sector: Soviet scientists and engineers discovered Siberia's enormous hydrocarbon fields, developed the world's largest oil and gas industry, and built (at the time of their completion) the world's longest pipelines, which supplied much of Europe's crude oil and natural gas needs.

But by 1991, when the country unraveled—remarkably, without any violence—the USSR was suffering from many innovation gaps, ranging from those in key primary industries to those needed to satisfy basic consumer demand. Steel is the dominant metal of modern civilization, and by the early 1990s no open-hearth furnaces were being used in the EU, North America, and Japan to make it—basic oxygen furnaces had begun to displace them in the 1950s—but this nineteenth-century process (introduced to make steel during the 1860s) was still used in the last years of the USSR to make nearly half of the USSR's metal output. And the country's lagging innovation in the mass production of common consumer items, ranging from blue jeans to personal computers, was among the perennial causes of public discontent and, undoubtedly, a factor contributing to the Soviet regime's demise.

In contrast to Soviet innovation failures, the post-1990 economic development of China is the best recent, and historically unequaled, example of mass-scale innovation based on rapid appropriation of a wide array of foreign inventions. The Chinese economy has not grown fourteen-fold in size and the country's average per capita income has not grown more than eleven-fold (both measured in constant monies) because of an unprecedented flow of transformative domestic inventions but because of the mass-scale deployment of devices or practices mastered abroad decades (or years, for the latest advances) earlier and transferred to a newly receptive setting. Determined domestic efforts and trillions of dollars of foreign direct investment were accompanied by an enormous transfer of the latest machines, designs, and procedures. This has taken place by patent acquisition and by know-how shared by American, European, and Japanese companies eager to enter the Chinese market, and these legal transfers have been accompanied by wide-ranging and relentless industrial espionage.

The Chinese Communist Party learned the lesson from the USSR's disintegration well: no loosening of control similar to Gorbachev's attempt to reform an unreformable political regime but, in its scale, a truly unprecedented innovation-led economic expansion that resulted in rapid quality-of-life gains and left the party even more firmly in control. The very first commercial transaction after Richard Nixon's "China-opening" visit in February 1972 was the purchase of the world's most advanced ammonia synthesis plants, designed by America's M. W. Kellogg Company; the acquisition was critical in preventing another large famine in a country with a rapidly growing population and no modern fertilizer industry.

Subsequently, thousands of foreign companies (led by the largest multinationals, including Toyota, Hitachi, Nippon Steel, GM, Ford, Boeing, Intel, Siemens, and Daimler) shared their know-how with China, typically by being forced into joint ventures that provided complete know-how for Chinese reverse engineering. All too obviously, China has benefited from being a late starter riding a huge innovation wave generated by the adoption of perfected foreign inventions. Of course, Japan and South Korea also traveled that road, starting respectively in the 1950s and the 1970s, but along the way they became not just determined innovating powers but also important inventive economies. Notable examples of these contributions range from Sony's lead in the early development of consumer electronics and Toyota's low-fault, just-in-time factory management to the development of advanced microprocessors, mobile telephony, and batteries (by, among others, Samsung, SK Hynix, LG, and Panasonic). So far there have not been any comparably important, globally embraced, and commercially rewarding Chinese contributions (although some might argue that Huawei should be included).

In looking back at the long trajectory of inventions, it is hardly surprising that many historians and economists have been impressed by the acceleration of these advances. Separating the frequency and consequences of truly epochal nineteenth-century inventions from the much less intensive and much more gradual technical progress of the eighteenth century is the Industrial Revolution. But the advances of the twentieth century were perhaps even more remarkable. As Joel Mokyr has pointed out, they took place despite two protracted world wars and despite the rise of totalitarian regimes that extended their rule over much of Europe and Asia:

> In the past, such catastrophes might have been enough to set economies back for hundreds of years or even to condemn entire societies to stagnation or barbarism. Yet none of them could stop the power of ever-faster innovation in the twentieth century to stimulate rapid growth in much of the industrialized and industrializing world.

The notion of ever-faster innovation ranks high among the incessantly recited mantras of the late twentieth and early twenty-first centuries. Obviously, a rising number of patents is not a perfect measure of this innovative acceleration (too many patents protect minor variations and marginal improvements on influential discoveries), but it is undeniable that the decadal aggregates of applications granted by the US Patent and Trademark Office (USPTO), including grants to foreign residents, increased from just 911 during the first decade of the nineteenth century to nearly 250,000 during the 1890s, and then went from about 340,000 during the first decade of the twentieth century to about 1,653,000 during the 1990s, a nearly 2,000-fold increase in two hundred years.

Of course, this simple, unqualified, and in some ways obviously misleading ascent of the total number of patents has always included dubious entries and even some truly mad creations. In 1932 Alford Brown and Harry Jeffcot put together a small collection of such cases from the files of the USPTO. One must wonder what possessed professional patent evaluators to grant protection to such items as an "improved burial-case" (whereby a person can "on recovery to consciousness, ascend from the grave and the coffin by the ladder") or a "device for producing dimples." If you think that we have left such frivolities behind, checking regularly the Electronic Frontier Foundation's web page "Stupid Patent of the Month" will make it clear that there is no shortage of such idiocies.

I would single out US patent 8,609,158B2, granted in 2013, and a lengthy quotation is necessary to make it clear how dubious the patenting process remains. The patent granted to a single inventor, Diane Elizabeth Brooks, is for Diane's manna,

> a potent drug with narcotic benefits made from distinctly and uniquely combined and processed interchangeable seed and seed derivatives that are so potent that it removes or alleviates depression, mood disorders, Attention Disorder

symptoms, thought disorder, mental illness, pain, right lip retardation symptoms, physical problems, Lymph Node cancer and many other illness symptoms. It removes bumps in the neck within a week or two. It is interchangeable in most aspects. . . . It is extremely strong or potent and can be made weak to make your little Attention deficit child normal. It is an incredible mood stabilizer and reduces psychosis. Use it for cancer patients and for people with pain issues. It works."

It boggles the mind that this claim was actually approved, but there are also many thoroughly factual grants that are still in the shake-your-head category, among them US patent D670,286S1 granted in 2012 to Apple (there were ten applicants, including Steve Jobs and the company's chief designer, Jonathan Ive) for a "portable display device," that is, for a rectangle with rounded corners (fig. 1.2). I cannot resist citing yet another American patent application by Susan R. Harsh for "a kit and method that converts dog nose smudges deposited on a first surface into a form of dog nose art on a second surface." Remarkably, this one has yet to be granted.

There are actually some revealing ways of evaluating patterns and identifying real breakthrough inventions (and I will introduce them in this book's closing chapter), but right now we can simply point out actual quantitative and qualitative improvements that took place thanks to what many believe is an accelerating flow of inventions—and then see those accomplishments not as completions but as mere foundations for further, and accelerating, progress. Modern inventions thus carry the promise of brilliant salvations as they are to solve every problem we face, technical, environmental, or social. Moreover, the solutions are promised to arrive not just as some marginal or gradual advances but as changes best described by such adjectives as "disruptive," "transformative," or "revolutionary"—and their nearly imminent world-changing potential is to extend to everything from food to longevity and from energy to travel.

We have already reduced the number of malnourished people to less than a tenth of the global population, so why not eliminate food shortages entirely—and, while we are at it, why not sever our dependence on field crops by producing food in climate-regulated high-rises or by swallowing synthetic capsules providing complete nutrition? During the past two centuries we have already doubled the average life expectancy in

Figure 1.2 The third image in Apple's US patent D670,286S1 application (issued in November 2012) showing a "portable display device"—by now a well-known rectangular shape with rounded corners. *Source:* J. Akana et al., Portable display device (US Patent D670,286S1, filed November 23, 2010, and issued November 6, 2012), https://patents.google.com/patent/USD670286.

affluent countries, so why not at least double it again through ingenious gene manipulation, or CRISPR our way to immortality? And during the same time, rich countries have multiplied (at different rates) the useful per capita supply of energy, so why not keep expanding it even as we eliminate all fossil carbon as an energy source by ingenious conversions of renewable sources of energy? We can already routinely travel at speeds of around 300 km/h on land and at nearly at the speed of sound (close to 1,000 km/h) in the air, so why not travel at supersonic speeds in buried or elevated vacuum tubes or in passenger planes that cross the Atlantic in a couple of hours?

And given the exponential (ever-faster) pace of modern invention, we are repeatedly told there is nothing extraordinarily audacious or unrealistically ambitious about such goals. The math is unavoidable: it is an inevitable property of long-lasting exponential growth that it ends up in a singularity, a point in time when a function reaches an infinite value, making anything instantly possible. But one does not have to be a disciple of the approaching Singularity cult because even relatively much more mundane claims are impressive—and keep on coming, announcing breakthroughs in treating diseases (drugs that supposedly cure Alzheimer's disease), storing electric energy (the invention of batteries of unheard-of energy density), and even converting other planets into habitable worlds (terraforming Mars). The realities have been far less exalted, and this book is a modest reminder of the world as it is, not the world of exaggerated claims or, even worse, the imaginary world of indefensible fantasies.

Before I go any further, I must note that I am not concerned here with the numerous design failures that resulted in catastrophic events (including such widely known tragedies as *Titanic*'s 1912 sinking and the *Challenger*'s 1986 launch disaster), in famously missing the commercial boat (Sony's Betamax video cassette recording device eliminated by JVC's VHS), or in notorious embarrassments (Ford's Edsel and Pinto, Google's Glass). Historians of technical advances have detailed many of these failures in studies dealing with such hopeless designs as electric ploughs in pre–World War I Germany or Chrysler's automotive gas turbines. And a recent listing reviews Apple's twelve most embarrassing product failures, from Macintosh TV to the Power Mac G4 Cube.

Those interested in this failed design genre should consult Susan Herring's 1989 book, *From the Titanic to the Challenger,* which lists no fewer than 1,354 such failures during the twentieth century, or Michael Schiffer's *Spectacular Flops,* where they can read about some older examples (including Tesla's World System of wireless electricity distribution) and some recent delusions (a nuclear reactor–powered bomber). At the same time, it must be appreciated that many design failures of engineered objects and systems are not only inevitable but offer great lessons (albeit often costly, sometime tragic) about what to avoid and what to correct;

that is why Henry Petroski subtitled his book devoted to these experiences *The Role of Failure in Successful Design*.

Similarly, this book is not about the many undesirable, often troublesome, and sometimes even fatal consequences of many eagerly accepted, massively propagated, and safely established modern inventions. These side effects, downsides, and complications have often been anticipated; many of them have been closely monitored, evaluated, and translated into monetary and quality-of-life costs, and they have also been the subjects of much research and efforts to prevent or mitigate them. The health and environmental impacts of prescription drugs are perhaps the most widely appreciated category of side effects in modern societies. They range from discomfort to strict contraindications dictated by preexisting conditions and from the presence of drug metabolites in streams and water bodies to the spread of antibiotic-resistant bacteria. The last is a very serious and now also a global problem. We have known about its advancing impact for many decades, but despite repeated exhortations and promises the search for new antibiotics still receives only a fraction of the resources and commitment that it deserves.

No less remarkable has been the tolerance of the multiple side effects created by the invention of cars powered by internal combustion engines. Those engines gave us mobility, convenience, and the proverbial freedom of the road—but also harmful emissions, reordered cityscapes (rarely for the better), and a fatality frequency whose equivalent would not be tolerated for any widely used prescription medicine. Even in the most affluent countries we began to reduce the emissions (with catalytic converters, a new invention, coming to the rescue) only during the 1970s, but we still have no effective, widely adopted solutions for cars as part of sensible urban design, and the annual toll of vehicular accidents has recently been 1.35 million deaths of drivers and pedestrians.

These consequences of major inventions and our remarkably selective tolerance of their undesirable impacts and side effects could be extended to topics ranging from the intensive use of synthetic nitrogen fertilizers to land and water contamination by many types of plastics—and they would need a lengthy book for even a cursory coverage. Here I will adopt a more general approach to inventive failures by focusing on the fact that the flow of fundamental and enormously successful inventions that have created modern civilization during the past 150 years has been accompanied by a frustrating lack of progress in many key areas, as well as on the innovations that, to put it charitably, did not do as well as initially expected. In this book I examine three notable categories of these innovation failures: unfulfilled promises, disappointments, and eventual rejections.

I am aware that some historians of technical advances find the very term "failed technology" misleading because it seems to suggest (as Tom Carroll argued at the 1989 symposium on failed innovations) a positivist kind of linear reading of a momentum "that a potential innovation either has or does not have," whereas the more important distinction is to recognize that "success" or "failure" is a consequence of social choice. Undoubtedly, technical advances are not autonomous and are strongly influenced by social conditions and contexts—but, all too obviously, major influences go in the other direction, and it is often not in the power of open societies (or even the rulers in dictatorial states) to decide what innovation to embrace or to reject.

I start with inventions that were diligently sought, generally (and often enthusiastically) praised when they eventually arrived, rapidly commercialized, and embraced on a global scale. But eventually, even decades later, they turned out be so undesirable and convincingly so harmful both to humans and to the environment that they came to be viewed with widespread suspicion, and subsequently they were banned outright for the uses for which they were originally invented. The introduction of leaded gasoline enabled the smooth operation of internal combustion engines, but it took several decades before the resulting emissions of a neurotoxic heavy metal were widely recognized as an unacceptable trade-off and, starting with the US in 1970, countries began to ban the use of this additive. A ban on DDT applications as a widespread means of insect control began shortly afterward, and in 1987 a global agreement outlined the timetable for the gradual abandonment of chlorofluorocarbons, commonly used as refrigerants, whose rising atmospheric concentration was linked to the decline of stratospheric ozone.

The next category of failed inventions I consider includes three important examples of those advances whose initial promise appeared to ensure the eventual domination of their respective market niches: airships for affordable long-distance air transport, nuclear fission for electricity generation, and supersonic aircraft for speedy intercontinental travel. These innovations were commercialized and more or less widely deployed, but it did not take long to realize that they would not reach their initially hoped-for potential. Chronologically, airships were the first practical application to fail, and they did so spectacularly, as the *Hindenburg* in flames became one of the most reproduced images of a technical catastrophe. But that accident did not end airship dreams, and attempts at resurrecting this form of transport have continued even after jetliners rapidly conquered global aviation after 1960, and new proposals for better airships have appeared during the first two decades of the twenty-first century.

Nuclear fission is a case of missed expectations on a much grander scale, and it has been undoubtedly the foremost example of the phenomenon I call successful failure. Despite its considerable commercial deployment (with more than four hundred reactors operating on three continents) and despite its major contribution to electricity generation in several affluent countries, its current share of the global market remains far below what was expected of this complex technique in the early phases of its enthusiastic adoption: nothing else but total domination by the end of the twentieth century! The history of supersonic flight bears some resemblance to both these cases: for a time more successful than the use of airships, ultimately unable to compete but repeatedly resurrected by new designs whose proponents maintain (as do companies pushing new reactor designs) that this time will be different as the faster-than-sound airplanes will be able to conquer a viable niche in the global market.

The final examples illustrate in some detail the failure of expectations. I focus on three prominent examples of many highly desirable innovations whose mass-scale commercialization would be truly transformative and whose imminent success has been promised for generations but whose effective and affordable realizations always appear to be beyond the discernible horizon. The idea of high-speed travel in a vacuum (or, more likely, inside tubes with air pressure lowered to a small fraction of the atmospheric normal) has been around for more than two hundred years, and its recent, highly publicized resurrection under the misleading label of hyperloop offers an excellent opportunity to explain how this generations-old dream still waits for practical, convenient, reliable, and profitable commercialization.

My second example of a promised invention we are still waiting for belongs to a vastly less publicized category of needed advances, yet its arrival would be one of the most consequential achievements in history. If the world's staple grain crops (wheat, rice, corn, sorghum) were able—much like such leguminous grains as beans, soybeans, lentils, and peas—to supply a significant part of their nitrogen demand through symbiosis with nitrogen-fixing bacteria, we would not only increase the global grain harvests but be able to reduce the output and application of synthetic fertilizers, thereby saving a great deal of energy and preventing several categories of environmental pollution. And my final example is the commercial exploitation of nuclear fusion for electricity generation, a feat first promised by some leading physicists during the 1940s. This has been perhaps the most famous and definitely the most publicized example in the category of failing expectations, and I explain the remarkable persistence of this dream whose realization seems to always be just beyond the horizon.

Of course, every one of these three categories of innovative failure may be expanded by introducing other notable examples. In looking at the inventions that turned from welcome to undesirable, I could have added the story of hydrogenated oils, whose commercial success began in 1911 with the partial hydrogenation of cottonseed oil, giving Procter & Gamble its Crisco (crystallized cottonseed oil), fat that remains solid at room temperature. The use of trans fats (solidified oils) was expanded to an array of inexpensive butter and lard substitutes that had a long shelf life and made great baked goods and became a common choice for deep frying—until dietary research linked them to increased blood cholesterol levels and a higher risk of heart disease, and governments moved to control their everyday use.

When recounting inventions that were set to dominate but never reached that level of importance, I could have traced the rise and fall of Blackberry, the mobile phone of CEOs and presidents known for its security features and apparently destined to dominate the corporate world. But its prominence lasted only about ten years: its first smartphone was released in 2002, but by 2013 the company could not compete, and entered a protracted slide. And to any discussion of the inventions that we keep waiting for, the story of the hydrogen economy, perhaps the ultimate but ever-postponed solution to the increasingly pressing need for global decarbonization, would make an excellent addition.

A lengthy interesting book could be written about inventions that dominated their particular production or consumption sectors for generations, indeed, for more than a century, before they rather rapidly either completely disappeared, or were retained only as marginal curios kept alive by eccentric devotees, or were economically marginalized. The already mentioned open-hearth furnaces are perhaps the best example in the first category: between the 1870s and the early 1950s all primary steel was made by reducing the carbon level of cast iron from blast furnaces in these large vessels. Then, within a generation, they nearly disappeared in Japan and Europe, lingered a bit in North America, and some of these nineteenth-century artifacts survived into the twenty-first century (fig. 1.3). A fundamental transportation shift provides an example of an even faster retreat. Ocean liners dominated intercontinental passenger transport for nearly a century before they disappeared within only about a decade after the introduction of scheduled transatlantic jet flights.

And, of course, all older readers of this book have witnessed how the new world of microelectronics has created many examples of the rapid near demise and marginal survival of formerly admirable inventions whose services dominated globally for more than a century. Typewriters were displaced by personal computers and later also by portable electronics, cameras were replaced by smartphones, and physical modes of recorded music (records, tapes, compact discs) displaced each other before direct digital access marginalized them all. To be sure, typewriters, cameras, and vinyl recordings are still around, but typewriters can be acquired only secondhand by those who prefer the mechanical option for writing, the market for cameras with exchangeable lenses is now overwhelmingly restricted to professional photographers and serious practitioners of, most often, nature photography, and recorded music is a nostalgic marginal niche in a world dominated by streaming.

The final chapter opens with comments on the exaggerated reporting of new inventions. Uncritical media reports about breakthroughs and epochal beginnings, often under naively or ridiculously phrased

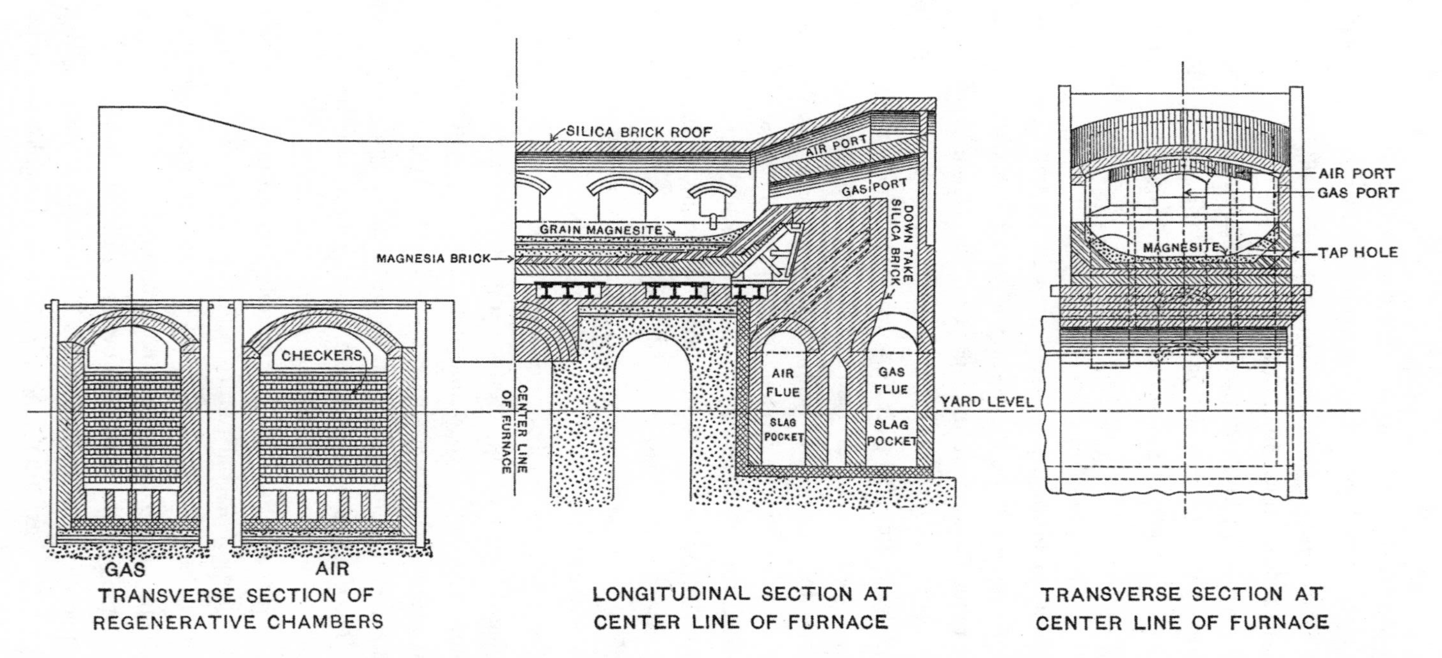

Figure 1.3 Sections through an early twentieth-century open-hearth furnace. *Source:* Harbison-Walker Refractories, *A Study of the Open Hearth* (Pittsburgh: Harbison-Walker Refractories, 1909). The last open-hearth furnace in the US was shut down in 1992, in China in 2001, and in Russia in 2018.

headlines, have become the norm that generates false conclusions and raises unwarranted expectations. This way of reporting has become so common that I review only some of the most egregious recent instances. Afterward, I contrast the now common belief in an ever-faster pace of innovation with the many unmistakable signs of technical stagnation and slowing advances: there are limits to everything, and invention and innovation cannot be exceptions. Consequently, there are no adulatory pages devoted to recent forecasts dwelling on the coming mastery of artificial intelligence (resulting in everything electronic, from autonomous vehicles to pilotless airplanes, and in machines making us irrelevant) or the creation of new life forms at will (genetic engineering unleashed on everything from pests to the human brain).

Obviously, we need many inventions whose large-scale adoption would provide long-overdue means to tackle some of our most daunting health, environmental, and economic challenges, from the conquest of malaria to reducing the (now actually widening) global income disparities. The book ends with a brief wish list of some much-needed advances. As in the past, we will succeed in some quests but fail in others, and we will not be able to ignore the fact that many gains will take place within limits rather than being the products of unlimited progress. We should restrain our ever-present compulsion to forecast how new inventions will shape our future: retrospectives of such efforts show only very limited success and a preponderance of failures. A better, safer, more equitable world will require many truly transformative inventions, but we will know the extent or absence of these expectations only when looking back—and we must hope that some of the items on my wish list will become realities before the middle of the twenty-first century.

2

INVENTIONS THAT TURNED FROM WELCOME TO UNDESIRABLE

Every solution of a complex problem, every helpful advance that eases or eliminates a specific harmful or undesirable impact, every innovation promising better performance, higher profits, or improved handling, or increased comfort or safety, has its obverse. Its reach and intensity range from predictable, tolerable, manageable (or simply time-limited) side effects to unforeseen yet potentially serious consequences that are not easy to deal with. Some of them can be eliminated only by abandoning the original solution in favor of a superior (entirely harmless) approach, or, if that is impossible, then at least replacing it with a less offensive, somewhat more acceptable choice.

I have chosen what I believe are the three most prominent examples of what eventually turned out to be unacceptable solutions to important, common, and, if they were to remain unaddressed, harmful and costly problems. All three of these innovations appeared during the interwar period—two of them deployed compounds known for decades, tetraethyl lead and dichlorodiphenyltrichloroethane, while one exploited a newly discovered halogenated compound, dichlorodifluoromethane—and I address them chronologically. First comes the introduction of leaded gasoline (starting in the US in 1922) as an inexpensive, convenient, and effective solution to the problem of suboptimal internal combustion engine operation widely known as knocking, premature ignition that not only reduced the machine's energy conversion efficiency but that could cause serious damage to the engine itself.

One of the most incredible coincidences in the history of innovation is that Thomas Midgley, the same engineer who headed the corporate search for an effective antiknocking agent that ended up with leaded

Figure 2.1 Thomas Midgley Jr. (1889–1944), the inventor of leaded gasoline and chlorofluorocarbon refrigerants. Portrait from the 1930s by Blank & Stoller, New York. *Source:* Williams Haynes Portrait Collection (Philadelphia, Science History Institute), box 10. https://digital.sciencehistory.org/works/9s161624t.

gasoline, would, just a few years later (in 1928), lead a group of researchers that formulated a nontoxic and nonflammable dichlorodifluoromethane (CCl_2F_2), sold under the brand name of Freon-12 (fig. 2.1). This was the first of many chlorofluorocarbons (CFCs), synthetic compounds that rapidly became the world's dominant refrigerants (liquids used in the compression-expansion cycle in refrigerators and air conditioners) and were also used as common blowing agents in the production of foams, as propellants in billions of aerosol cans (containing medicine, paints or cosmetics), and as industrial degreasing agents and solvents.

The final example of a much-welcomed innovation that turned into a much-maligned application is DDT (dichlorodiphenyltrichloroethane), the first modern synthetic insecticide. By the time Paul Hermann Müller began his search for a powerful agent able to kill common insect pests, DDT had been known for more than six decades, but only his systematic quest for an effective agent led to the discovery of the compound's formidable insecticidal power. DDT found almost instant application by the armies of World War II, and after the war its use spread rapidly to control insect-borne infectious diseases, and for more general pest control in crop production and among livestock. In little more than a decade these uncontrolled practices not only led to the emergence of DDT-resistant insect species, they were also linked to adverse effects on bird reproduction and eventually also to higher risks of premature birth weights or of low-birth-weight babies, and DDT became one of the destructive symbols used by the nascent environmental movement to spread its message of more responsible management.

Besides their common trajectories of ascent and demise, leaded gasoline, CFCs, and DDT have had their specific routes of acceptance and elimination. When lead was first added to gasoline there was plenty of convincing evidence of its insidious neurotoxicity, and the new product was met with near-instant resistance by a number of physicians and physiologists. In contrast, Freon-12 was a new synthetic compound that did not exist in nature and that, fortuitously, appeared to be quite unreactive when accidentally released into the environment, making it a perfect choice for household refrigerants. Midgley might be criticized for his role in introducing tetraethyl lead as the dominant antiknocking agent, but to say, as Neil Larsen did, that he was "the most harmful inventor in history" is nonsense.

In 1928 it could have been anticipated that CFCs released into the atmosphere would, though much heavier than air, eventually reach the stratosphere: turbulent atmospheric mixing does the same for CO_2, the leading greenhouse gas that is also heavier than air. But it was only half a century later that advances in the study of atmospheric chemistry made it clear that chlorine is released from CFCs during dark polar winters owing to reactions taking place on the surfaces of icy particles, and that when the Sun returns the photochemical reactions of the freed element with the stratospheric ozone start reducing concentrations of the gas that provides indispensable protection against ultraviolet radiation. Similarly, there was no previous experience with DDT because prior to its deployment we had only such natural insecticides as citrus and eucalyptus oils or water solutions of salts or neem oil (extracted from the seeds of a tropical evergreen tree, *Azadirachta indica*), and even if the first toxicological studies of the early 1940s had been done far more extensively and rigorously, they would not have uncovered the long-term cumulative effects on bird reproduction.

And the trajectories also differed in their lengths, and in their final phases. Eight decades elapsed between the introduction of leaded gasoline and the complete global ban on its use, with Indonesia being the last nation that allowed its sale until 2006. The first identification of CFCs as potential destroyers of stratospheric ozone was published in 1974, forty-six years after Freon-12 formulation, and in 1987 the Montreal Protocol on Substances That Deplete the Ozone Layer outlined the steps leading to the complete global ban on the use of CFCs. DDT reached the peak of its global application in only about two decades after its introduction. Actions to limit and outlaw its use began during the 1960s, and the compound is now banned worldwide, with the exception of regulated use to control malarial mosquitoes.

The most encouraging lesson common to the history of these three notable failures has been our ability not only to come up with better alternatives but also to devise practical international arrangements to make the bans and substitutions effective (with some notable breaches) on a global scale. With gasoline we had such options even before the unfortunate selection of lead as the most convenient additive, and the eventual elimination of the heavy metal was indefensibly delayed. In contrast, actions to reduce and eventually to eliminate CFCs as the cause of stratospheric ozone reduction proceeded swiftly and resulted in one of the most effective global treaties. The consequences of the DDT ban are much harder to assess because the compound's introduction was followed by the formulation of scores of other pesticides (not only insecticides but also compounds combating worms and fungi), and health and environmental impact studies have indicated that the chronic application of many of them is hardly risk-free.

There is yet another disquieting commonality. These three innovations were products of targeted corporate research (General Motors looking for an antiknocking compounds and better refrigerants, and Swiss Geigy searching for effective insecticides). Their commercialization required approval by regulating bodies, but this requirement was not able to prevent the introduction of potentially dangerous environmental contaminants. Not only was tetraethyl lead approved despite well-known risks and against the clearly voiced objections of leading health scientists, but its use persevered for a lifetime, and its eventual ban was not (or at least not primarily) the consequence of belated concerns about its neurotoxicity. And both CFCs and DDT were initially welcomed not only as near-perfect solutions to technical problems but also as innovations delivering major health benefits, namely, eliminating toxic (and potentially deadly) refrigerants, particularly in household settings, and eradicating the common insect vectors of (potentially even more deadly) diseases.

The history of tetraethyl lead is, in the first place, the story of failed public health measures: if the known risks had been taken into account, there would not have been, decades later, a failed invention and the need to ban the compound's use. CFCs and DDT carry different, much more sobering but also expected lessons: human interventions in Earth's environment often carry delayed, complex risks, so far removed from the initial concern and so far beyond the readily conceivable complications that only time and the accumulation of events will make us aware of those unexpected but highly consequential impacts. Extraordinary diligence, commitment, and imagination should reduce the scope of such delayed revelations, but it is highly improbable that their recurrence can be completely eliminated.

LEADED GASOLINE

The mass adoption of road vehicles powered by internal combustion engines—in 2022 the world had more than 1.4 billion of them on the road—is a perfect example of a highly complex fundamental innovation resulting from combinations of advances in the design and manufacturing of internal combustion engines, primary metals (steel, aluminum, nickel, vanadium), tires (rubber), and electrical components (batteries, switches, starters), which were integrated through improvements in machine optimization and manufacturing (the moving assembly line) and enabled by the development of reliable fuel sources (crude oil extraction and refining) and essential infrastructure (paved roads, pipelines, filling stations).

Consequently, the question of who invented motor vehicles cannot have a simple answer. In 1886 Gottlieb Daimler and Wilhelm Maybach mounted a water-cooled engine on a wooden coach and, independently,

Karl Benz put a light single-cylinder engine on a three-wheel chassis. But the only component these first drivable machines—tall, open, slow, and clumsy-looking—had in common with today's road vehicles was their (much less powerful and much less efficient) internal combustion engine. All else, from wheels to steering, from chassis to the placement of engines, has changed profoundly. It took the remainder of the nineteenth century and the combination of innovations contributed by German, French, British, and American engineers to turn these awkward hybrid designs, initially looking like horseless carriages, into the real precursors of modern cars. In 1901 the Maybach-designed Mercedes 35 was the first essentially modern motor vehicle: still without any roof but including four cylinders, two carburetors, mechanical inlet valves, an aluminum engine block, a gear stick in a gate, a honeycomb radiator, and rubber tires.

Further advances followed: just seven years later Henry Ford began to sell his Model T, the first mass-produced affordable and durable passenger car, and in 1911 Charles Kettering, who later played a key role in developing leaded gasoline, designed the first practical electric starter, which obviated dangerous hand cranking (fig. 2.2). And although hard-topped roads were still in short supply even in the eastern part of the US, their construction began to accelerate, with the country's paved highway length more than doubling between 1905 and 1920. No less important, decades of crude oil discoveries accompanied by advances in refining provided the liquid fuels needed for the expansion of the new transportation, and in 1913 Standard Oil of Indiana introduced William Burton's thermal cracking of crude oil, the process that increased gasoline yield while reducing the share of volatile compounds that make up the bulk of natural gasolines.

But having more affordable and more reliable cars, more paved roads, and a dependable supply of appropriate fuel still left a problem inherent in the combustion cycle used by car engines: the propensity to violent knocking (pinging). In a perfectly operating gasoline engine, gas combustion is initiated solely by a timed spark at the top of the combustion chamber and the resulting flame front moves uniformly across the cylinder volume. Knocking is caused by spontaneous ignitions (small explosions, mini-detonations) taking place in the remaining gases before they are reached by the flame front initiated by sparking. Knocking creates

Figure 2.2 Charles F. Kettering (1876–1958), inventor of the first practical electric starter, longtime (1920–1947) head of research at General Motors, and the man who insisted on calling the leaded additive "ethyl gas."

high pressures (up to 18 MPa, or nearly up to 180 times the normal atmospheric level), and the resulting shock waves, traveling at speeds greater than sound, vibrate the combustion chamber walls and produce the telling sounds of a knocking, malfunctioning engine.

Knocking sounds alarming at any speed, but when an engine operates at a high load it can be very destructive. Severe knocking can cause brutal irreparable engine damage, including cylinder head erosion, broken piston rings, and melted pistons; and any knocking reduces an engine's efficiency and releases more pollutants; in particular, it results in higher nitrogen oxide emissions. The capacity to resist knocking—that is, fuel's stability—is based on the pressure at which fuel will spontaneously ignite and has been universally measured in octane numbers, which are usually displayed by filling stations in bold black numbers on a yellow background.

Octane (C_8H_{18}) is one of the alkanes (hydrocarbons with the general formula C_nH_{2n+2}) that form anywhere between 10 to 40 percent of light crude oils, and one of its isomers (compounds with the same number of carbon and hydrogen atoms but with a different molecular structure), 2,2,4-trimethypentane (iso-octane), was taken as the maximum (100 percent) on the octane rating scale because the compound completely prevents any knocking. The higher the octane rating of gasoline, the more resistant the fuel is to knocking, and engines can operate more efficiently with higher compression ratios. North American refiners now offer three octane grades, regular gasoline (87), midgrade fuel (89), and premium fuel mixes (91–93).

During the first two decades of the twentieth century, the earliest phase of automotive expansion, there were three options to minimize or eliminate destructive knocking. The first one was to keep the compression ratios of internal combustion engines relatively low, below 4.3:1: Ford's bestselling Model T, rolled out in 1908, had a compression ratio of 3.98:1. The second one was to develop smaller but more efficient engines running on better fuel, and the third one was to use additives that would prevent the uncontrolled ignition. Keeping compression ratios low meant wasting fuel, and the reduced engine efficiency was of a particular concern during the years of rapid post–World War I economic expansion as rising car ownership of more powerful and more spacious cars led to concerns about the long-term adequacy of domestic crude oil supplies and the growing dependence on imports. Consequently, additives offered the easiest way out: they would allow using lower-quality fuel in more powerful engines operating more efficiently with higher compression ratios.

During the first two decades of the twentieth century there was considerable interest in ethanol (ethyl alcohol, C_2H_6O or CH_3CH_2OH), both as a car fuel and as a gasoline additive. Numerous tests proved that engines using pure ethanol would never knock, and ethanol blends with kerosene and gasoline were tried in Europe and in the US. Ethanol's well-known proponents included Alexander Graham Bell, Elihu Thomson, and Henry Ford (although Ford did not, as many sources erroneously claim, design the Model T to run on ethanol or to be a dual-fuel vehicle; it was to be fueled by gasoline); Charles Kettering considered it to be the fuel of the future.

But three disadvantages complicated ethanol's large-scale adoption: it was more expensive than gasoline, it was not available in volumes sufficient to meet the rising demand for automotive fuel, and increasing its supply, even only if it were used as the dominant additive, would have claimed significant shares of crop production. At that time there were no affordable, direct ways to produce the fuel on a large scale from abundant cellulosic waste such as wood or straw: cellulose had first to be hydrolyzed by sulfuric acid and the resulting sugars were then fermented. That is why the fuel ethanol was made mostly from the same food crops that were used to make (in much smaller volumes) alcohol for drinking and medicinal and industrial uses.

The search for a new, effective additive began in 1916 in Charles Kettering's Dayton Research Laboratories with Thomas Midgley, a young (born in 1889) mechanical engineer, in charge of this effort. In July 1918 a report prepared in collaboration with the US Army and the US Bureau of Mines listed ethyl alcohol, benzene, and a cyclohexane as the compounds that did not produce any knocking in high-compression engines. In 1919, when Kettering was hired by GM to head its new research division, he defined the challenge as one of averting a looming fuel shortage: the US domestic crude oil supply was expected to be gone in fifteen years, and "if we could successfully raise the compression of our motors . . . we could double the mileage and thereby lengthen this period to 30 years." Kettering saw two routes toward that goal, by using a high-volume additive (ethanol or, as tests showed, fuel with 40 percent benzene that eliminated any knocking) or a low-percentage alternative, akin to but better than the 1 percent iodine solution that was accidentally discovered in 1919 to have the same effect.

In early 1921 Kettering learned about Victor Lehner's synthesis of selenium oxychloride at the University of Wisconsin. Tests showed it to be a highly effective but, as expected, also a highly corrosive antiknocking compound, but they led directly to considering compounds of other elements in group 16 of the periodic table: both diethyl selenide and diethyl telluride showed even better antiknocking properties, but the latter compound was poisonous when inhaled or absorbed through skin and had a powerful garlicky smell. Tetraethyl tin was the next compound found to be modestly effective, and on December 9, 1921, a solution of 1 percent tetraethyl lead (TEL)—$(C_2H_5)_4$ Pb—produced no knock in the test engine, and soon was found to be effective even when added in concentrations as low as 0.04 percent by volume.

TEL was originally synthesized in Germany by Karl Jacob Löwig in 1853 and had no previous commercial use. In January 1922, DuPont and Standard Oil of New Jersey were contracted to produce TEL, and by February 1923 the new fuel (with the additive mixed into the gasoline at pumps by means of simple devices called ethylizers) became available to the public in a small number of filling stations. Even as the commitment to TEL was going ahead, Midgley and Kettering conceded that "unquestionably alcohol is the fuel of the future," and estimates showed that a 20 percent blend of ethanol and gasoline needed in 1920 could be supplied by using only about 9 percent of the country's grain and sugar crops while providing an additional market for US farmers. And during the interwar period many European and some tropical countries used blends of 10–25 percent ethanol (made from surplus food crops and paper mill wastes) and gasoline, admittedly for relatively small markets as the pre–World War II ownership of family cars in Europe was only a fraction of the US mean.

Other known alternatives included vapor-phase cracked refinery liquids, benzene blends, and gasoline from naphthenic crudes (containing little or no wax). Why did GM, well aware of these realities, decide not only to pursue just the TEL route but also to claim (despite its own correct understanding) that there were no available alternatives: "So far as we know at the present time, tetraethyl lead is the only material available which can bring about these results"? Several factors help to explain the choice. The ethanol route would have required a mass-scale development of a new industry dedicated to an automotive fuel additive that could not be controlled by GM. Moreover, as already noted, the preferable option, producing ethanol from cellulosic waste (crop residues, wood) rather than from food crops, was too expensive to be practical. In fact, the large-scale production of cellulosic ethanol by new enzymatic conversions, promised to be of epoch-making importance in the twenty-first century, has failed its expectations, and by 2020 high-volume US production of ethanol (used as an antiknocking additive) continued to be based on fermenting corn: in 2020 it claimed almost exactly one-third of the country's corn harvest.

In contrast, Midgley's TEL patent—titled, unhelpfully, "Method and means for using motor fuels"—filed on April 15, 1922 (and issued on February 23, 1926), gave the company full control of an effective low-volume additive that could be dispensed at a very low cost: a penny's worth of TEL would prevent knocking from consuming a gallon of gasoline (fig. 2.3). Worst of all, and truly unpardonable, has been the denial of any possible health concerns. This effort began with Kettering's insistence on the inaccurate naming of the additive ("ethyl gas"), which deliberately avoided acknowledging the presence of lead. This heavy metal, known for its toxicity since Greek antiquity, was sometimes claimed to have played a major role in the demise of the Roman Empire, and by the early twentieth century it was well known as a cause of health problems associated with various occupational exposures. But GM and its TEL suppliers were not just engaged in disregarding lead's health effects, they made resolute and repeated claims aimed at minimizing or even entirely dismissing any concerns about the health effects of a compound to be emitted into the environment from car exhaust on such a large scale.

The ancient understanding of lead's toxicity advanced considerably during the nineteenth century with the clear identification of chronic lead poisoning leaving serious neurotoxic damage, with unborn children and infants being particularly vulnerable. Not surprisingly, some of America's leading public health experts opposed the addition of lead to gasoline and asked for an investigation of likely dangers. GM and DuPont claimed, without doing any studies, that the average street would likely be so free from lead that it would be impossible to detect its absorption. But in late October 1924, thirty-five workers in the TEL processing plant in New Jersey experienced acute neurological symptoms, and five of them died. By coincidence, the Bureau of Mines released its TEL investigation on the day the last victim of acute exposure died; the report concluded there were no dangers to the general public. This was immediately criticized by several leading physiologists, and on May 20, 1925, the US surgeon general, responding to public concerns, convened a conference in Washington, D.C., to confront the contending claims.

At this meeting GM, DuPont, Standard Oil, and Ethyl Corporation framed the use of TEL as a necessity required to ensure the industrial progress of the country. Frank Howard of the Ethyl Corporation stated, "Our continued development of motor fuels is essential in our civilization," and saw TEL's discovery to be "an apparent gift of God" to conserve oil. Such claims were strongly rejected, with Alice Hamilton, a physician at

Feb. 23, 1926. 1,573,846

T. MIDGLEY, JR

METHOD AND MEANS FOR USING MOTOR FUELS

Filed April 15, 1922

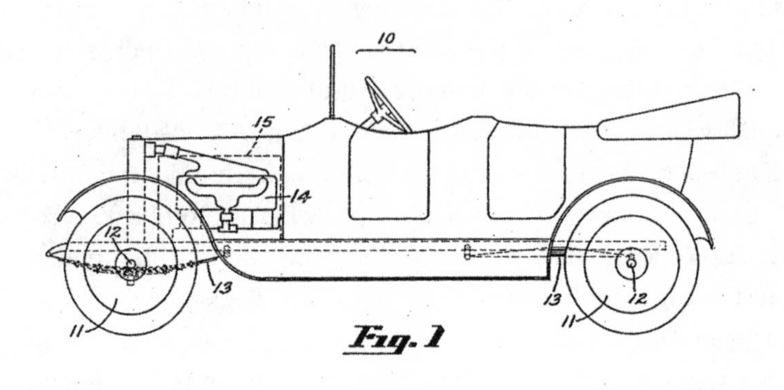

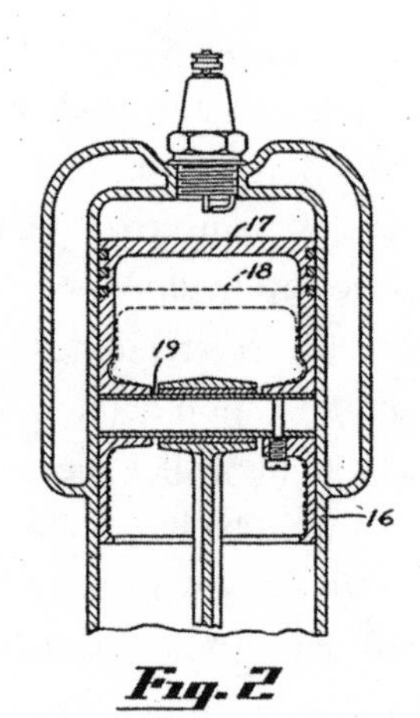

Witnesses
Irwin A. Greenwald
Lloyd. M. Keighley

Inventor
Thomas Midgley Jr.

By Francis D. Hardesty
Attorney

Figure 2.3 Midgley's curiously named and no less curiously illustrated patent application for the use of leaded gasoline in automobiles. *Source:* T. Midgley Jr., Method and means for using motor fuels (US Patent 1,573,846, filed April 15, 1922, and issued February 23, 1926), https://patents.google.com/patent/US1573846.

Harvard Medical School, pointing out "that lead is a slow and cumulative poison and that it does not usually produce striking symptoms that are easily recognized," and concluding that "I am not one of those who believe that the use of this leaded gasoline can ever be made safe. No lead industry has ever, even under the strictest control, lost all of its dangers." The conference ended with Ethyl Corporation's announcement that it was suspending the production and distribution of leaded gasoline pending the results of an independent investigation. But this apparent victory of TEL's opponents was just a delay and a detour on the road to the mass adoption of leaded gasoline.

The promised study began in October 1925 in Ohio and enrolled just 252 workers, divided into four groups. The controls included thirty-six men who drove cars and twenty-one garage workers or filling station attendants who did not come into contact with leaded gasoline, while the persons at risk were seventy-seven chauffeurs and fifty-seven filling station attendants exposed to TEL gasoline, as well as sixty-one men in factories known to have lead dust exposure. All too obviously, a study allowed to run for just seven months from its design to its final report was quite inadequate to uncover the long-term effects of exposure to lead, and the final report submitted to the surgeon general in May 1926 concluded that "there are no good grounds for prohibiting the use of ethyl gasoline of the composition specified as a motor fuel, provided that the distribution and use are controlled by proper regulation," but it also called for further studies: "The committee feels that this investigation must not be allowed to lapse." But lapse it did, better studies never took place, and the notion that human progress could not continue under what the industry's leaders considered onerous restrictions had prevailed, a course undoubtedly aided by the country's post-1929 economic hardships.

In 1927 the office of the surgeon general set a voluntary standard for adding no more than 3 grams of TEL per gallon of gasoline. The American standard for producing leaded gasoline was gradually adopted around the world, allowing for a doubling of the compression ratio (typically to 8.3–10.5:1) and greater efficiency of car engines. Besides saving energy in driving, tetraethyl lead in aviation fuel made it possible to develop more powerful, faster, and more reliable reciprocating aeroengines, machines that reached the peak of their performance during World War II before they were eclipsed by gas turbines. And after the war, when the US intensified the interrupted automobilization and car ownership began to spread in Europe and Japan, the production of leaded gasoline reached new heights. These developments were used to justify the original claims of TEL promoters who saw the additive as a fundamental breakthrough for the US car industry that ensured its economic might and, until the 1970s, its global dominance.

Remarkably, in 1958 the surgeon general's office actually raised the maximum allowable TEL addition to 4.23 grams per gallon (g/gal; because it had no indication of rising levels of lead in blood or urine), while the actual industry average during the 1950s and 1960s was about 2.4 g/gal. During the three decades between 1945 and 1975 the US consumed nearly two trillion gallons of gasoline, which means (using the 2.4 g/gal average) that it added about 4.7 million tons of lead to the environment via vehicle exhausts, with annual additions surpassing 200,000 tons a year during the early 1970s. Meanwhile, advances in toxicology made it clear that serious health consequences were not limited to relatively high acute or chronic occupational exposures. During the 1940s it became clear that lead causes retardation in growth, behavioral disorders, and intellectual impairment in children, and starting in the 1970s we had realized that these effects arise even from "silent" doses, that is, from relatively low, prolonged asymptomatic exposures, all avoidable by banning the use of lead-containing compounds.

The first major source of these exposures was lead in household paints, added as lead oxide, carbonate, or chromate in order to resist moisture, increase durability, and speed up drying. Its danger was recognized at the beginning of the twentieth century, but only in 1977 were lead-containing paints banned in the US, and their use was permitted even longer in Europe and Asia. Lead in gasoline was a much larger source of toxic environmental pollution, but this massive use and resulting environmental contamination raised little or no concern during the 1950s (recall the raised maximum lead allowance) and 1960s, and it was only in 1970, after forty-four years of globally increasing lead emissions from TEL gasoline, that the US finally began the process of removing the toxic metal from the most important refined liquid fuel—and health concerns were not the decisive reason for this shift.

By that time the country's large cities were experiencing repeated, and often protracted, periods of photochemical smog, an air pollution phenomenon that arises from complex atmospheric reactions of carbon monoxide, nitrogen oxides, and volatile hydrocarbons emitted from the refining, distribution, and combustion of liquid fuels. Photochemical smog, first noted in Los Angeles in the 1940s, eventually became a seasonal presence in all large metropolitan regions. The US Clean Air Act of 1970 gave the newly formed Environmental Protection Agency (EPA) power to regulate harmful compounds, and in 1973 the agency mandated major reductions in automotive emissions and the phased removal of lead from all grades of gasoline.

A technical fix for photochemical smog became possible in 1962 when Eugène Jules Houdry patented a way to remove the pollutants from vehicle exhaust just before their emission into the atmosphere by deploying catalytic converters. Platinum was used as the rare metal catalyst; it would be poisoned by lead's presence in exhaust gases, and this made the introduction of effective catalytic converters (mandatory in all cars starting with the 1975 model year) dependent on the availability of unleaded gas. Eventually these devices made a decisive difference as the precontrol emissions of hydrocarbons and carbon monoxide were cut by 96 percent and those of nitrogen oxides by 90 percent.

In 1970 unleaded gasoline only had about 3 percent of the US market. By 1975 this figure had increased to 12 percent, and starting in 1979 the US EPA required all refineries to lower the average lead content in leaded fuels: it fell to just 1 g/gal by 1980, 0.5 g/gal by 1985, and 0.1 g/gal by 1988. At the same time, increased awareness of the health costs of exposures to lead—with studies showing adverse impacts on IQ in children and on hypertension in adults—accelerated the complete phase-out of leaded fuel. In 1985 unleaded gasoline had 63 percent of the market; by 1991 it had 95 percent. In 1985 an EPA study estimated the value of the benefits of the final lead phasedown (effects on children, reduction in other pollutants, improvements in maintenance) to be at least twice that of the associated costs (higher refinery expenditures) and twelve times that when the costs of adult male hypertension were added.

Some measurable effects appeared soon: as the lead phase-out proceeded, the median lead concentration in American children decreased by nearly 80 percent between 1976 and 1994, and by 2015 it was only about 5 percent of the mid-1970s level. A recent study led by Anna Aizer has shown that even further reductions in lead from historically low levels have significant positive effects on children's third-grade reading test scores: every unit decrease in average blood lead levels reduced the probability of a child being substantially below proficient in reading by about 3 percent. Other countries followed the US lead. Japan banned leaded fuel by 1986, but in Europe the use of leaded fuel began to decline only in 1986 in Germany, 1988 in France, and 1990 in Spain. The EU finally banned leaded fuel in 2000, the same year as China and India. The two penultimate holdouts were Venezuela, which initiated a ban in 2005, and Indonesia, which did so in 2006, and it was not until July 2021 that Algeria stopped selling leaded gasoline.

What has displaced TEL? Methyl tertiary butyl ether (MTBE) became a leading additive by the late 1990s, but in 2000 the EPA announced its phase-out because of its adverse environmental effects (its solubility in water led to contamination of aquifers). This left refiners with two major choices: either reformulate gasolines with increased shares of hydrocarbons that prevent knocking (known as the BTEX complex) or turn to ethanol. Initially, the BTEX complex became the leading substitute: this mixture of hydrocarbons—benzene, toluene, ethyl benzene, and xylene—present in liquid fuels is separated by refining and added to gasoline (which contains a limited volume of these aromatics) to boost its antiknocking capacity. Remarkably, the efficacy of benzene blends was already well known, and even used in some US markets, when GM began its TEL push in 1925!

As the substitution progressed, the average BTEX share rose from 22 percent to 33 percent of volume by 1990, and up to 50 percent in premium gasolines. This led to new health concerns, and the EPA eventually set the BTEX limit at 25–28 percent of gasoline volume, but concerns about the mixture's health effects remain. Fortunately, there are no worrisome adverse effects caused by burning a mixture of gasoline and ethanol, and crop-derived ethanol (in the US overwhelmingly from corn, in Brazil from sugar cane) became the leading antiknocking additive. The rise of US ethanol began in earnest in 2005 when the Energy Policy Act set the minimum volumes of biofuels to be blended with transportation fuels, and in 2020 blends of 90 percent gasoline and 10 percent ethanol (known as E10) accounted for more than 95 percent of all fuel used by the country's gasoline vehicles.

Unfortunately, I cannot close with even a rough quantitative contrast of benefits and costs, only with a few incontrovertible observations. The mass-scale introduction of tetraethyl lead during the mid-1920s provided a quick and dirty solution to an important technical problem, and because it enabled higher engine efficiencies, it did have its environmental benefits: everything else being equal, it resulted in lower relative emission rates (grams/kilometer); but much larger car fleets composed of heavier cars erased these relative gains, and total emissions of all car-related pollutants kept on increasing until the mid-1970s. The invention's value was in its simplicity, ready applicability, and low cost, not in any unprecedented brilliance, and most definitely, TEL was not the only option to conquer engine knocking.

The invention's peril, evident since the very beginning but hidden under the misleading label of ethyl gasoline, had its worst cumulative effect on children exposed to lead from car exhaust, in the US for six decades between the mid-1920s and the mid-1980s, in the rest of the world mostly during the second half of the twentieth century. We have identified many facets of the cumulative exposure to low levels of lead among children: lower scores on general intelligence tests and on reading; compromised visuospatial functions, memory, attention span, processing speed, and language ability; and effects on motor skills (manual dexterity) and affective behavior. Moreover, research did not find any threshold below which lead remains without effect on the central nervous system, and a 1993 study by the National Academy of Sciences confirmed that lead causes neurobehavioral deficits even in extremely low doses.

Hence the most tragic consequence of the tetraethyl lead used as a fuel additive was the differential reduction of equal chances for success in life owing to chronic childhood exposures to a neurotoxin. These exposures may not have shortened the overall life span but they deprived millions of children of an equal chance for a successful life. Obviously, lead in car exhaust was just one of several unavoidable deprivations experienced disproportionately by children of lower socioeconomic status, but because of its undoubted neurotoxic impact it cannot be dismissed as marginal and nonconsequential. It is impossible to quantify the cumulative impacts of these exposures across generations and on a global scale, but it is hard to avoid the conclusion that few inventions that were initially extolled as perfect solutions to a technical problem caused so much avoidable deprivation on the individual level.

And how to explain that great puzzle of American society, giving the leading car industry and oil companies generations of carte blanche despite the known risks of exposure to lead, and the opposition's failure to come back after the initial defeat? Did the combination of chronic, unseen, and insidious exposures simply pale in comparison to the concerns arising from the unprecedented economic crisis of the 1930s, the global war of the early 1940s, the rush to prosperity and the Cold War of the 1950s and 1960s? Would we still have leaded gasoline had we not had to reduce intolerable levels of photochemical smog and prevent the metal poisoning platinum catalyzers?

DDT

Killing insects has never been an easy matter: their size, their often seasonally exploding ubiquity, their fitness (resulting from their long evolution, starting some 400 million years ago), and, for flying species, their elusive three-dimensional mobility make any large-scale complete eradication unattainable, and even on smaller scales keeping their numbers in check requires repeated and expensive control measures. Consequently it is rather surprising that a deliberate, systematic search for compounds whose efficacy would far surpass the relatively modest and time-limited effects of known natural insecticides began only during the late 1930s.

Paul Hermann Müller got his doctorate in organic chemistry in Basel in 1925 and became employed by the research division of J. R. Geigy, a dye-making company whose origins go back to the mid-eighteenth century. Müller's first assignment was to work on synthetic and plant-derived dyes and tanning agents. A decade later he moved on to the synthesis of plant protection compounds (mothproofing for textiles) and developed new products with bactericidal and insecticidal properties as well as Graminone, a new seed disinfectant that replaced mercury-based compounds. His next assignment was to come up with new insecticides that would replace expensive (and often only marginally effective) low-potency natural products or affordable but toxic arsenic compounds: the eventual rise of a new compound to global prominence began with yet another corporate search for better alternatives.

At that time, the outlook for better insecticides was not good. The ideal compound should have a rapid toxic effect on as many species as possible, but it should have no (or minimal) toxicity for mammals or plants and be nonirritating and nonodorous, persistent (chemically stable) and affordable. None of the insecticides known at that time—including, most commonly, pyrethrum (extracted from chrysanthemum flowers and imported mostly from Japan), rotenone (present in some legumes), and nicotine (found in tobacco and other nightshade plants)—had persistent activity, most of them were expensive, and some of them were toxic or an irritant to people.

In 1939, after four years of research and the testing of 349 possible candidates, Müller found a promising molecule. Experiments done by others in his company showed that compounds with a chloromethyl (-CH_2Cl) group had oral toxicity to moths, and he became aware of a 1934 paper in the *Journal of the Chemical Society* in which two British authors described the preparation of a diphenyltrichloroethane and was curious to know whether the -CCl_3 group had any insecticidal activity on contact. Accordingly, in September 1939 he synthesized ***d***ichloro***d***iphenyl***t***richloroethane, and tests immediately showed that it had an insecticidal contact effect unmatched by any known compound.

But it was not an unknown molecule: this organochlorine compound was first synthesized in 1874 by Othmar Zeidler, an Austrian chemist, when he studied at the University of Strasbourg (at that time part of Germany after France's defeat in 1871). Zeidler did nothing to exploit any practical applicability of his discovery, not an unusual course of action during the second half of the nineteenth century: those decades abounded with new syntheses of organic compounds that found no immediate use, including (as already noted) tetraethyl lead, an organometallic compound synthesized first in 1853, and polyvinylchloride (PVC), now the world's second most important plastic (only polyethylene is produced in larger amounts) that was synthesized by Eugen Baumann in 1872.

Müller's identification of DDT—a colorless, tasteless, nearly odorless crystalline compound—as a potent insecticide was quickly confirmed by further tests showing it killed mosquitoes, lice, fleas, sandflies, and Colorado potato beetles. Patents followed quickly (Swiss 1940, British 1942, American 1943), and Geigy began to distribute the insecticide in two concentrations containing 5 percent (Gesarol spray against the potato beetle) and 3 percent (Neocide dust, primarily to control lice) DDT. Thanks to the intervention of the US military attaché in Bern, samples were received in New York in November 1942, and the US military, running short on pyrethrum, began to use the compound to fight malaria, typhus, and lice, first in Europe and then on the Pacific islands.

The results were convincing. During two summer months of 1943 in Sicily, the US Army had 21,482 hospital admissions for malaria compared to 17,375 battle casualties (wounded and dead). A public health poster had it right: "The malaria mosquito knocks out more men than the enemy." Field testing of DDT began in Italy in August 1943; by 1945 new cases of malaria had declined by more than 80 percent, and DDT was also in use, in an indiscriminate but highly effective fashion, to stop the typhus epidemic in Naples. Starting in mid-December of 1943 some 1.3 million people were dusted (they had to tie their clothes at the wrists and ankles, and DDT powder was dispensed down their collars and waists), and two months later the city had no new typhus cases. By the end of World War II and its immediate aftermath, DDT was widely used by Allied armies, in evacuations of concentration camps and prisons and in the repatriation of deportees.

This record of rapid and highly effective disease prevention gave DDT a very positive public image that was only strengthened by postwar efforts to eradicate malaria, first in the US and in parts of southern Europe. In 1948 Paul Müller was awarded the Nobel Prize in Physiology or Medicine "for his discovery of the high efficiency of DDT as a contact poison against several arthropods," with the citation concluding that "without any doubt, the material has already preserved the life and health of hundreds of thousands" (fig. 2.4). And that total number of saved lives kept on growing: in 1970 the National Academy of Sciences' Committee on Research in the Life Sciences concluded that "to only a few chemicals does man owe such a great debt as to DDT" because in less than two

Figure 2.4 Paul Hermann Müller won the 1948 Nobel Prize in Physiology or Medicine for his work on DDT.

decades of use, it had prevented 500 million deaths from malaria, and the compound became one of the new tools (besides the new short-stalked, high-yielding wheat and rice cultivars receiving increasing amounts of synthetic nitrogenous fertilizers) in the global quest to eradicate hunger, malnutrition, and diseases.

In the US, DDT became available for sale to the public in October 1945 as both an agricultural and a household pesticide, and its application in crop protection grew rapidly. Of course, in light of the compound's manifest neurotoxicity to insects, there were concerns about its health effects. One of the earliest appraisals was published in 1945 by Patrick Buxton, a leading British entomologist, who concluded that DDT combined high insecticidal power with low toxicity to mammals, and that while large doses could cause pathological changes in the liver and tremors, there was no evidence of harm to people who made it or applied it. Hence, "After 2 years of very wide experience, I feel that we may say that, used as an insecticide, DDT is harmless." But there were concerns about DDT's persistence: clothes impregnated with it could kill lice even after several washings, and films of DDT deposited on walls or glass panes would keep killing mosquitoes and flies for many weeks.

The first reports of adverse effects came during the late 1950s as both agricultural applications and large-scale DDT spraying to control mosquitoes, tent caterpillars, and gypsy moths became common. In 1958 Derek Ratcliffe at the British Nature Conservancy reported his findings of the relatively sudden appearance of abnormally large numbers of broken eggs in eyries of peregrine falcons during the early 1950s. In the same year, Roy Barker of the Illinois State Natural History Survey published a paper in the *Journal of Wildlife Management* calling attention to *"the possibility that moderate applications of DDT under certain conditions can be concentrated by earthworms to produce a lethal effect on robins nearly one year later."* This was based on the effect of the spring spraying of elm trees on the main campus of the University of Illinois in Urbana with a 6 percent solution of DDT between May 1950 and May 1952: during that time, twenty-one dying robins were found on the campus, all with elevated levels of DDT or its metabolite in their brain. This finding became the key part of an extended indictment of DDT that was published four years later.

In the same year when Ratcliffe and Barker published their findings, Rachel Carson—a marine biologist formerly employed by the US Fish and Wildlife Service who left her job in 1952 after her previous publication, *The Sea Around US,* became a bestseller that gave her financial independence—began to investigate the anti-DDT activities among some communities, mostly in the US Northeast, affected by DDT spraying (fig. 2.5). These groups set up the Committee Against Mass Poisoning and in one case even filed an injunction against the US Department of Agriculture. Carson also began to collect information about the risks of pesticides, originally intending to report on these matters in an article for the *New Yorker*. But with more information forthcoming, Carson decided to write a book instead. The manuscript was first edited (and abridged) for serialized appearance in the *New Yorker* starting in June 1962, then published in a massive edition by Houghton Mifflin and selected as the Book-of-the-Month Club choice, and finally turned into a televised CBS presentation.

This trifecta turned the book into the best-known nonfiction work of the 1960s. *Silent Spring* presented the use of DDT as one of the most consequential human interferences in the natural order of things, and the book was intended to make the widest possible public impact. Its title

Figure 2.5 Rachel Carson (1907–1964), whose book *Silent Spring* helped turn the American public against DDT. *Source:* US Fish and Wildlife Service.

refers to a letter written in 1958 by a resident of Hinsdale, Illinois, following several years of spraying of elms by DDT:

The town is almost devoid of robins and starlings; chickadees have not been on my shelf for two years, and this year the cardinals are gone too; the nesting population in the neighborhood seems to consist of one dove pair and perhaps one catbird family. It is hard to explain to the children that the birds have been killed off. . . . "Will they ever come back?" they ask, and I do not have the answer.

Hence the evocative image of silent springs descending on America.

Looking ahead, Carson offered some frightening scenarios. In the book's opening "Fable for Tomorrow," Carson deliberately conflated realistic possibilities with absolutely unsupportable exaggerations of children dying almost instantly:

The farmers spoke of much illness among their families. In the town the doctors had become more and more puzzled by new kinds of sickness appearing among their patients. There had been several sudden and unexplained deaths, not only among adults but even among children, who would be stricken suddenly while at play and die within a few hours.

> There was a strange stillness. The birds, for example—where had they gone? . . . It was a spring without voices.

And the book's eleventh chapter, dealing with the toxicity of pesticides, unsubtly described the makers of pesticides as poisoners "beyond the dream of the Borgias."

Although *Silent Spring* took a broader view of human impact on the biosphere, and although Carson had repeatedly pointed out how other pesticides, whose formulation and commercialization followed DDT's introduction, were much more toxic and far more harmful to biota (*"endrin . . . makes the progenitor of all this group of insecticides, DDT, seem by comparison almost harmless"*), she made DDT the centerpiece of the book's lengthy indictment: throughout the book Carson refers to it nearly two hundred times. What followed its publication was covered only with the use of superlatives.

The book became an immediate bestseller (and remained so on the *New York Times'* list for eighty-six weeks). It was seen as an unprecedented indictment, a "shattering tsunami" of revelations that "launched the modern environmental movement," much as Harriet Beecher Stowe's *Uncle Tom's Cabin* had engendered antipathy to slavery and Thomas Paine's *Common Sense* had summed up the radical sentiment at the outset of the American Revolution. Inevitably, books have been written about this book, and it has become one of those publications whose message became clear even to very large numbers people who never read it (and to generations born after its publication who heard about it): DDT kills and harms in many ways.

As the book led to widespread support for banning DDT's use, further investigations detailed newly identified harmful effects, above all the role of the compound in the catastrophic declines of raptorial birds, especially peregrine falcons and bald eagles, in parts of the US, and the confirmation that DDT was definitely more harmful to robins than their exposure to methoxychlor, another commonly used insecticide that was banned only in 2003. In 1971 and 1972 the newly established US EPA held seven months of hearings on DDT, resulting in more than nine thousand pages of transcribed testimony, and Edmund Sweeney, the EPA's hearing examiner, issued a 113-page report of recommended findings, conclusions, and orders that was published in the *Federal Register* in April 25, 1972. He found that DDT should not be banned because it had essential uses; it was not the sole offender in the family of pesticides (with some replacements having more deleterious effects); it was "not a carcinogenic, mutagenic, or teratogenic hazard to man"; and the uses under regulations "do not have deleterious effect on fresh water fish, estuarine organisms, wild birds, or other wildlife."

But just six weeks later the agency's administrator, William Ruckelshaus, issued the decision to ban the compound, basing it on a combination of factors that "constitute a risk to the environment." The leading concerns justifying his decision were DDT's concentration in organisms (both terrestrial and marine), its transference through the food webs, its persistence in soil for years (even decades), its contamination of aquatic ecosystems, its lethality to many beneficial insects, its toxicity to fish, its role in thinning bird eggshells and hence impairing reproduction, and its possible carcinogenic risk. These factors "constitute an unknown, unquantifiable risk to man and lower organisms," and hence "an unacceptable risk" arising from any further use, and warranted a precautionary ban on its further use as insecticide applied to many common crops, including cotton, corn, beans, peanuts, and vegetables.

This decision, coming less than two years after a committee of the US National Academy of Sciences singled out humanity's debt to DDT, was strongly criticized, and not only by the companies involved in making and applying the insecticide. Among others who joined in disbelief was Norman Borlaug, whose high-yielding varieties greatly boosted the world's staple crop yields and who thought the ban was a terrible decision, and some of the country's leading entomologists. In the midst of the American debate on banning DDT, *Science* printed a letter by a University of California entomologist who saw the ban as a judgment of emotion and mystique and argued that there was no evidence of harm to people or animals from legitimate DDT uses despite its widespread, high-volume applications. Another biologist, this time from Rutgers University, wondered "how far this absurd campaign will go to replace effective, safe, and proven pesticide."

Later, after being accused of making a political decision, Ruckelshaus explained his reasoning:

I talked to a reporter from *Chemical Week* who asked whether it was political. I said small "p" political—in the sense of a society trying to decide what risk it's willing to accept for what benefits. But I'm not talking about big "P" politics. His editorial said that I admitted it was a political decision.

But the case against DDT does not rest either on Carson's imaginary exaggerations or on Ruckelshaus's decision to ignore the examiner's findings. The decision to ban DDT's use was justified by a deeper understanding provided by post-1972 studies, which made clear that the compound's environmental effects warranted the precautionary prohibition of most of its uses and that the ban did not, contrary to some claims, have any major regrettable consequences.

The US ban, preceded by the Swedish ban in 1970, did not end all uses of the chemical: exemptions could be granted, and during the 1970s DDT was used to suppress typhus- and plague-carrying fleas, weevils, and moths in several states, including Louisiana, California, Colorado, New Mexico, and Nevada. But as large-scale agricultural spraying ended, levels of DDT in biota (fatty tissues, blood) began to decline, and eggshell thinning caused by DDT and DDE (dichlorodiphenyldichloroethylene, DDT's metabolite) was eventually studied on all continents except Antarctica. The impairment shows significant interspecific variation: most notably, chicken and quail have not been affected at all, while raptors and fish-eating species have been most susceptible as a result of the bioaccumulation of DDT and DDE in fatty tissues.

Because the calcium concentration in blood of affected birds remains normal, it is most likely that DDE affects the mineral's transport across the eggshell gland mucosa, reducing the shell thickness by up to 50 percent (most commonly by 15–25 percent). Direct measurements of shell thickness of eggs from the pre-DDT era that had been preserved in museum collections were impossible without breaking them (the contents of the egg are removed through tiny holes, precluding the use of a micrometer), and Ratcliffe devised instead an index (weight/length × breadth) that made comparisons with newly collected eggs possible. And David Peakall, later at the National Wildlife Center in Ottawa, realized that, thanks to the persistence of DDE, he might be able to measure its content in the desiccated membrane that remained inside emptied eggs—and he was indeed able to fill decades-old eggs from museum collections with hexane, and chromatographic analyses revealed the presence of DDE.

His studies of British peregrine falcon eggs showed no traces of DDE from eggs collected in 1933, 1936, and 1946, but four of five clutches from 1947 did. As a result, by the early 1960s peregrine falcons had disappeared completely in Britain and throughout the eastern US and southern Canada. Other raptors affected by eggshell thinning included ospreys, bald eagles, sparrow hawks, red-tailed hawks, and, as recently as between 2006 and 2010, also condors that had been reintroduced in central California and were feeding on the carcasses of sea lions living in the Southern California Bight, which had been contaminated in the past by wastes discharged from a DDT factory. Fish-eating birds with documented losses have included double-crested cormorants, brown pelicans, African fish eagles, great blue herons, and the white-faced ibis. Moreover, the persistence of DDT means that some bird populations have yet to revert to normal eggshell thickness: gains have been steady among Greenland's peregrines for decades, but the return to pre-DDT normal may not take place until 2034.

The gradually declining presence of DDT and DDE and the captive breeding and reintroduction of raptors in the most affected areas have resulted in widespread recoveries of previously extirpated or much-reduced species. But how has the DDT ban affected the fight against malaria-carrying mosquitoes, the compound's most common agricultural application? There were initial rapid successes in Sardinia, Greece, and the southern United States, but during the 1950s, as more countries turned to large-scale DDT spraying, natural selection resulted in the emergence of DDT-resistant mosquitoes. As Morag Dagen has noted, "Mosquitoes had adapted to DDT before the planned worldwide antimalarial campaign had even begun."

The European and American DDT bans of the early 1970s did not apply elsewhere (and insecticide's production for export continued in the US until the mid-1980s), and India, the leading DDT user and exporter (mainly to Africa) kept on expanding its output: a new DDT-producing plant opened in Maharashtra in 1977 and another one in Punjab in 2003. But by that time India was only one of the three remaining producers (together with China and North Korea, the latter in tiny amounts). During the late 1990s negotiations began on a global agreement to eliminate the most offensive persistent organic pollutants. They were completed in 2001, and the Stockholm Convention became legally binding in May 2004: initially it outlawed nine compounds, including the insecticides aldrine, chlordane, endrin, lindane, and mirex, and limited the use of DDT to malaria control in tropical countries.

In 2006 the World Health Organization revisited its DDT guidelines and confirmed that the compound is the most effective of the twelve insecticides approved for indoor use (able to reduce malaria transmission by up to 90 percent) and that its correct applications do not harm either people or wildlife. In 2011 the World Health Organization reiterated that "DDT is still needed and used for disease vector control simply because there is no alternative of both equivalent efficacy and operational feasibility, especially for high-transmission areas," and stated that the reduction and ultimate elimination of its use depends on the development of alternatives and on financial help for the poorest countries. India remained the largest user during the second decade of the twenty-first century, and only in 2015 did it begin to negotiate its accession to the Stockholm Convention. By 2019, eleven countries, including India, Mexico, Brazil, and six in sub-Saharan Africa, still approved indoor residual spraying with DDT, and the failure to eradicate malaria globally cannot be ascribed to restricted DDT use.

By 2019 the disease was endemic in eighty-seven countries with about 230 million cases, but Africa's sub-Saharan countries accounted for 91 percent of all cases, and just two, Nigeria and the Democratic Republic Congo, accounted for nearly 40 percent. Resistance has been a part of this eradication failure. By the end of the twentieth century more than fifty species of anopheline mosquitoes had become DDT resistant, including those that are leading malaria vectors across sub-Saharan Africa and in Asia, and resistance to other compounds is also common. By 2019, seventy-three countries were reporting resistance to at least one insecticide in one malaria-transmitting species, and twenty-eight countries were reporting resistance to all four main insecticide classes. This resistance seriously weakens but does not negate the compound's capacity for mosquito control (its toxicity may be much lower, but it may still act as a repellent and an irritant). Despite some setbacks, the continued use of DDT during the 1950s and the 1960s eradicated malarial mosquitoes in North America, Europe, and much of the Caribbean. Why not in Africa?

As Michael Palmer has noted, the success of eradication does not hinge on the use of any individual compound but rather on the aggregate capacity to carry out multiple measures that prevent, control, and limit the disease, starting with economic development and including hygiene, surveillance, and treatment: the overlap between the high rates of new malaria infections and low levels of prosperity is all too obvious. DDT's role in malaria control remains contested: besides the always vocal proponents of a total DDT ban who minimize the compound's role in post-1945 eradication, there are still many pro-DDT advocates who see any limits on DDT use as counterproductive and who claim that DDT bans have resulted in the deaths of millions. There is a centrist DDT position, which recognizes that in some instances there is still no better control option than indoor residual spraying—but that an unqualified labeling of DDT as a safe choice for this application is untenable as it ignores accumulated evidence that argues for precaution.

No human populations were ever in danger (*pace* Carson) of seeing their children *"stricken suddenly while at play and die within a few hours,"* but more than seventy-five years after the beginning of DDT's large-scale applications we have a fairly good understanding of its health effects. We know that acute exposures to DDT produce a variety of responses ranging from heightened excitability, tremors, dizziness, and seizures to sweating, headache, nausea, and vomiting. Chronic occupational exposures can lead to permanent behavioral changes ranging from diminished attention and the loss of synchrony between visual information and physical movement to a variety of neuropsychological and psychiatric symptoms.

In 2008 a meeting convened to address the current and legacy implications of DDT production acknowledged both the benefits arising from the past prevention of insect-borne diseases and substantial exposures to DDT and DDE from continued indoor residual spraying, whose risks, compared to occupational exposures, have been rarely investigated. Moreover, as with so many other exposures, children, pregnant women, and immunocompromised individuals may be most at risk, and malaria-endemic areas with indoor DDT spraying also have high rates of HIV/

AIDS. Because of the lipophilic nature of DDT/DDE, prolonged breastfeeding in Africa may expose infants to undesirably high doses.

A 2019 evaluation of studies published during seven decades of research shows inconsistent evidence as far as most noncancer and cancer outcomes are concerned, with only some studies finding associations. Consistent evidence shows an association of DDT exposure with abortion or preterm births, with the prevalence of wheezing in infants and children, and with liver cancer. But even in these cases the links are observational, not causal; moreover, most of the studies claiming an association were done without controlling for exposure to other organochlorines that may be associated with DDT. Limited evidence links DDT to non-Hodgkin's lymphoma and to testicular cancers, and in 2015 the International Agency for Research on Cancer (IARC) classified DDT as "probably carcinogenic to humans." DDT may also suppress the immune system and acts as an endocrine disruptor, possibly raising the incidence of breast cancer.

This history of DDT use makes it clear that the trajectory of its rise and fall has some obvious similarity to the ascent and eventual demise of leaded gasoline but that the overall cumulative impact of the insecticide's large-scale and decades-long use is much harder to assess. Leaded gasoline did enable higher combustion efficiencies but there is little doubt that the health benefits resulting from reduced automotive emissions were far outweighed by the introduction of a known and persistent neurotoxin into the environment. In contrast, DDT's indisputably positive role in eliminating malaria from many countries and reducing its burdens in others could have been even more positive had we not resorted to massive spraying of crops, which burdened the environment with a persistent pollutant and led to the widespread rise of DDT/DDE tolerance among targeted insects.

In the end, DDT became just one of several persistent organic pesticides that had to go. The retreat began in Sweden in 1971 and, most significantly, in the US with the EPA's precautionary ruling in 1972, and by 2001 the compound headed the list of twelve chemicals tagged by the Stockholm Convention for complete or nearly complete elimination. Internal residual spraying is still going on in India and some African countries, but the trajectory of ascent (accompanied by marveling at the compound's lasting insecticidal powers) and decline (generated by DDT's effects on biota and by the rise of widespread resistance to it) is now nearly complete.

DDT now belongs to the category of inventions that were not just welcome but seen as truly transformative, only to be relegated to the class of undesirable advances. Could it have been different if, from the very start, its use had remained tightly restricted to closely controlled antimalarial measures and the compound had never been used for the large-scale spraying of crops? Perhaps, but the compound's initial use by armies during the last years of World War II (to suppress disease vectors) and its later rapid adoption as a key ingredient of the Green Revolution precluded such cautious and closely controlled applications. At least in that way, DDT became a victim of its early success.

CHLOROFLUOROCARBONS

Refrigeration and air conditioning are perfect examples of ubiquitous technologies that are essential for the perpetuation of modern civilization but that work in the background, taken for granted, commonly invisible, and producing, when run properly, only faint sounds and the desired degree of coolness. Compressors, the devices that enable cooling and freezing, work steadily, hidden inside metal boxes without attracting the attention of media ever eager to report on the advances of artificial intelligence or genetic engineering. Perhaps the closest analogy to compressors are transformers. They are even more numerous devices used to step up or step down voltages so that electricity can be transmitted across long distances (using ultra-high voltages up to 1,100 kV) or mobile phones can be operable (relying on batteries of less than 5 V) and that do so, unlike sometimes noisy compressors, always in complete silence.

Before the advent of modern refrigeration, the choices for keeping foodstuffs and drinks were limited to cutting, transporting, and storing ice (this became a substantial seasonal industry during the nineteenth century) or evaporating water from porous clay vessels, and interiors could be kept cool only by shading, thick walls, or a building design that produced a cooling chimney effect. In 1805 Oliver Evans proposed a closed-cycle refrigerating ether-based system, and in 1828 Jacob Perkins and Richard Trevithick came up with an air-cycle setup, but both of these designs remained on paper. The real breakthrough came in 1834 when Perkins patented a mechanical refrigeration machine that used a volatile fluid, ethyl ether, as the refrigerant. Every modern refrigeration system has the same four parts: compressor, condenser, expansion valve, and evaporator, and the Perkins cycle became the foundation of new industrial refrigeration projects.

In 1855 came the first ice-making plant, in Cleveland; in 1861 the first meat-freezing plant, in Sydney. Steam engines provided the first reliable means of powering compressors, and beginning in the 1880s electricity offered a much better (quiet, clean at the point of use) form of energy—but there were no perfect refrigerants, compounds whose compression and subsequent expansion and recompression runs the refrigerating or cooling cycles. In his historical survey, James Calm described the first generation of refrigerants used for the first hundred years of the industry, the 1830s to early 1930s, as "whatever worked" and was readily available: the lists included ethers, hydrocarbons ranging from light methane (CH_4), ethane (C_2H_6), and propane (C_3H_8) to heavier isobutane, propylene, pentane (C_5H_{12}), carbon dioxide (CO_2), ammonia (NH_3), sulfur dioxide (SO_2), ethyl chloride (CH_3CH_2Cl), methyl formate ($HCOOCH_3$), and carbon tetrachloride (CCl_4).

The ideal refrigerant should be nonflammable, nontoxic, and nonreactive: if it gets spilled from a broken duct or from a malfunctioning compressor it should not ignite or asphyxiate or poison people or combine with other compounds it may encounter. CO_2 is nontoxic and noncombustible, but, being heavier than air, it may accumulate near low-lying areas inside confined spaces, and by displacing oxygen it can cause asphyxiation. In the absence of good alternatives even some flammable gases were touted as quite acceptable. In 1922 an advertisement claimed that propane "is a neutral chemical" that is "neither deleterious nor obnoxious," and, if needed, "the engineer can work in its vapour without inconvenience." True, propane has low toxicity, but because it is heavier than air, any leakage of the gas in enclosed spaces will cause its near-floor accumulation and carry the risks of fire and explosion.

In 1860 Ferdinand Carré patented a refrigeration cycle using ammonia, and that compound, despite its own risks—corrosive to skin, eyes, and lungs, acutely toxic at 300 ppm—became the preferred refrigerant in large industrial systems and remains so even today because it produces the best net refrigerating effect (heat absorbed per unit of refrigerant from the refrigerated space). Ammonia is also flammable (in concentrations of 15–28 percent by volume in air), but its flammability is less than that of hydrocarbon refrigerants, and its low odor threshold (just 20 ppm) makes it easy to detect even without sensors. Even so, large cold-storage facilities experience accidental leakages and have developed elaborate sensing and control arrangements.

Obviously, none of these "natural" refrigerants—flammable hydrocarbons, corrosive ammonia, toxic sulfur dioxide—offered a safe and highly acceptable choice for household refrigerators. During the late 1920s this became an obvious barrier to mass-scale adoption of home refrigeration. The first models of small food refrigerators became available in the US just before World War I, but widespread diffusion also depended on the rate of electrification and the cost of electricity. By 1925 half of America's households were connected to the grid, and the price of electricity was falling. A better refrigerant was the only basic improvement that was needed to make refrigerators as common as radios.

Meanwhile, General Motors became the owner of the country's leading refrigerator-making company. In 1915 Alfred Mellows designed his first Frigerator, but in 1918, after he had sold only small number of the devices in Detroit, his company was bought by William Durant, the founder of GM, who then sold it to GM. But it was not a thriving business. The Frigidaire design used SO_2 as the refrigerant, and because of the obvious health hazard, household refrigerators were kept outside on porches and could not be installed in hospitals or restaurants (fig. 2.6). Enter, once again, Charles Kettering, the head of GM's Research Laboratories. He realized that "the refrigeration industry needs a new refrigerant if they expect to get anywhere," and, looking further ahead, he was also thinking about a huge market for refrigeration (and eventually air conditioning) in tropical countries, and about air conditioning in cars.

As with the electric starter and leaded gasoline, Kettering decided to find a solution through targeted research, and Thomas Midgley, who, after his work on leaded gasoline, spent years working on synthetic rubber at Cornell University, agreed to head the search for a superior

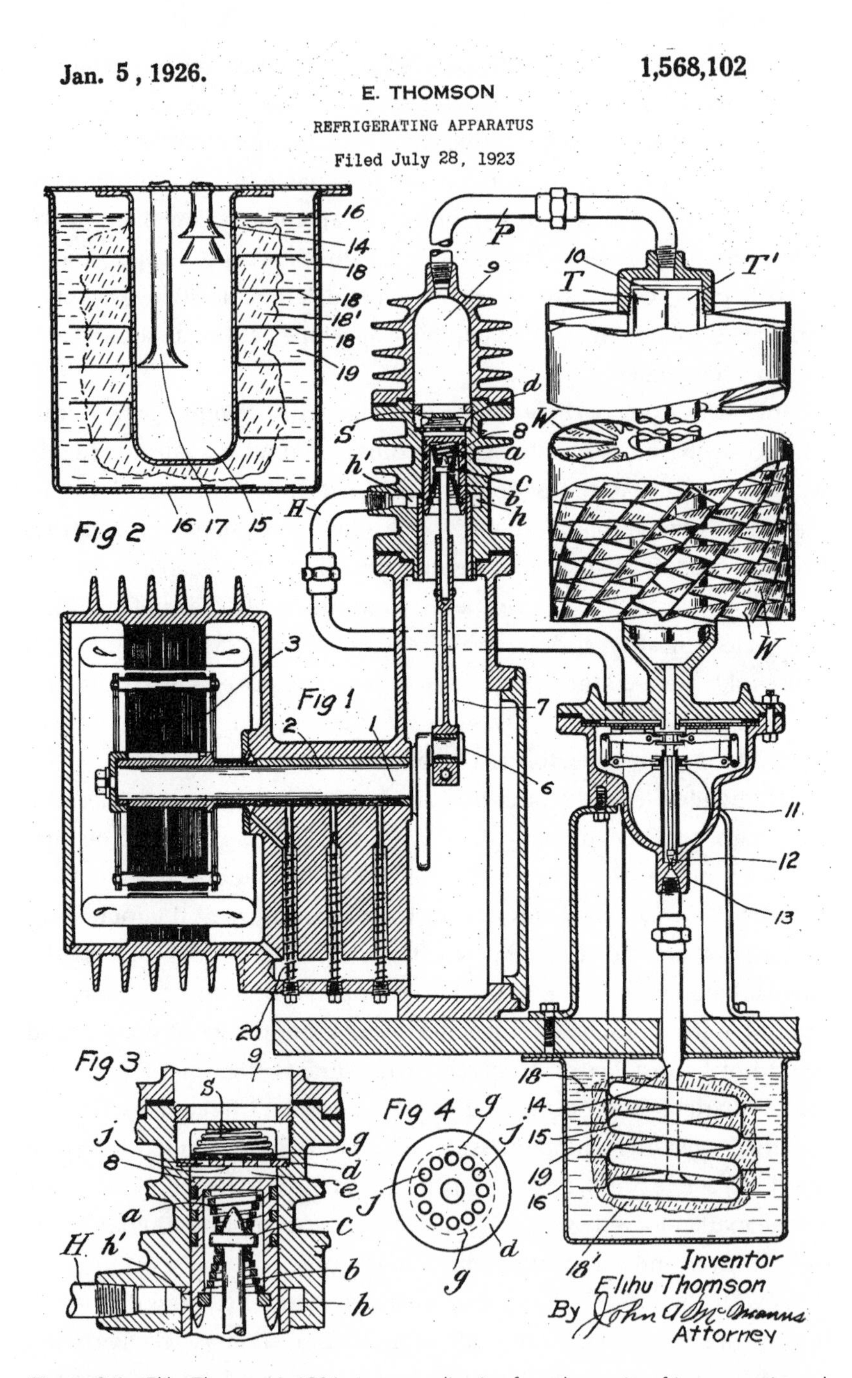

Figure 2.6 Elihu Thomson's 1926 patent application for a domestic refrigerator using sulfur dioxide. *Source:* E. Thomson, Refrigerating apparatus (US Patent 1,568,102, filed July 28, 1923, and issued January 5, 1926), https://patents.google.com/patent/US1568102A.

refrigerant. His closest research associates were Albert Henne, an expert on fluorine chemistry, who suggested that the element's substitutions in chlorinated compounds might produce a sought-after refrigerant, and Robert McNary. The first chlorofluorocarbon compound they synthesized was dichlorodifluoromethane (CCl_2F_2), known as F12 and sold under the proprietary name Freon, whose intermediate was trichlorofluoromethane ($CFCl_3$ known as F11), and although they did not make it, they were aware that they could also produce the overfluorinated alternative, chlorotrifluoromethane (CF_3Cl), known as F13.

They sniffed F12 and survived the experiment; then they organized a series of guinea pig tests proving the compound's safety. In April 1930 Midgley introduced Freon at the American Chemical Society meeting in a surprising manner, inhaling a bit of it on stage (nontoxic!) and slowly exhaling it to distinguish a candle flame (nonflammable!). In August 1930 GM and DuPont set up a joint stock company to make and market the compound, and Freon received its US patent (under the generic title *Heat transfers*) in November 1931. The business rewards were immediate. By 1929 GM had delivered its millionth refrigerator, by 1932 (even though the country was in the middle of the century's greatest economic recession) the total was up to 2.25 million units, and then, despite the continued economic crisis followed by World War II (with industrial mobilization for military production), the share of US households owning a refrigerator rose from just 10 percent in 1930 to nearly 60 percent in 1945 and to 90 percent in 1952.

This rapid diffusion of domestic refrigerators was then replicated in postwar Europe and Japan (in some countries the saturation was accomplished in little more than a single generation), and ownership of fridges began to spread to better-off urban families in low-income countries. By the early 1970s all affluent countries had more refrigerators than color TVs, and more Americans were also benefiting from two important applications of the Perkins cycle: by 1970 about half of all households had air conditioning, and so had more than half of new cars. And by that time household refrigeration and widespread space and car air conditioning were just two, though very important, uses of CFFs. The concatenation of desirable CFC properties—stable, noncorrosive, nonflammable, nontoxic, and affordable—also made them the ideal choices for aerosol propellants (used in products from cosmetics to paints, and medical inhalers), the production of plastic insulants (including polyurethanes, phenolics, and extruded polystyrene), the cleaning of delicate electronic circuits, and the extraction of edible and aromatic oils.

This resulted in an exponential rise in CFC production. The annual global output of the two dominant compounds, F-11 and F-12, later known as CFC-11 and CFC-12 or R-11 and R-12, rose from less than 550 tons in 1934 to more than 50,000 tons in 1950, to about 125,000 tons in 1960, and then it soared to the peak of 812,522 tons in 1974, with the US accounting for nearly half of the total and with America's DuPont and Allied Signal, Britain's ICI, and Europe's Akzo, Atochem, Hoechst, Kali-Chemie, and Montefluos being their major producers. And what was the fate of nearly 10 million tons of CFCs that had entered the atmosphere since the early 1930s? Nobody knew—until the first measurements of atmospheric concentrations of CFC-11, done exactly four decades after Midgley unveiled his team's discovery.

In 1970, James Lovelock, a British scientist best known for his Gaia theory, or Earth as a self-regulating superorganism, designed a procedure for measuring atmospheric levels of CFC-11. In 1971 he took the first readings in Adrigole in western Ireland and detected the compound not only during the easterly flows from the polluted European atmosphere but also in clean air coming from the Atlantic. He concluded that its presence "in the atmosphere is not in any sense a hazard," that the existence of the compound could be detected only by very sensitive electron absorption technique, and, because there are no natural gaseous fluorine compounds, its concentration could serve as an indicator of air masses contaminated by industrial pollutants.

Then in 1971 and 1972 Lovelock and his colleagues measured CFC-11 regularly on a ship traveling the Atlantic from England to Antarctica, finding the compound all along the traverse in an average concentration of around 50 parts per trillion, with (expectedly) higher levels in the Northern Hemisphere. The verdict was obvious: CFCs were staying in the atmosphere, and because of their inertia nearly their entire post-1930 output was accumulating aloft. But did the presence of these compounds, as Lovelock's group concluded, pose "no conceivable hazard" because they did "not disturb the environment"—or could their accumulation have undesirable consequences? Hypotheses suggesting the latter outcome were published in 1974. Richard Stolarski and Ralph Cicerone were the first ones to propose that chlorine oxides might be an important sink for stratospheric ozone, and showed how ozone molecules could be destroyed in two catalytic cycles.

Soon afterward Sherwood Rowland and his graduate student, Mario Molina, linked chlorine in CFCs directly to ozone destruction when they published a brief paper in *Nature* whose title explained the concern: "Stratospheric Sink for Chlorofluoromethanes: Chlorine Atom-Catalyzed Destruction of Ozone" and whose publication led, eleven years later, to a Nobel Prize in Chemistry. Atmospheric mixing eventually transports highly persistent CFCs into the stratosphere, and what happens there was succinctly summarized in Molina's Nobel lecture:

> The CFCs will not be destroyed by the common cleansing mechanisms that remove most pollutants from the atmosphere, such as rain, or oxidation by hydroxyl radicals. Instead, the CFCs will be decomposed by short wavelength solar ultraviolet radiation, but only after drifting to the upper stratosphere—above much of the ozone layer—which is where they will first encounter such radiation. Upon absorption of solar radiation the CFC molecules will rapidly release their chlorine atoms, which will then participate in the following catalytic reactions:
>
> $Cl + O_3 \rightarrow ClO + O_2$
>
> $ClO + O \rightarrow Cl + O_2$

Chlorine destroys ozone but then is released to start a new cycle of destruction, and a single atom of the gas can destroy on the order of 100,000 ozone molecules before it is eventually removed from the stratosphere by downward diffusion and reactions with methane. This was a highly worrisome hypothesis because the stratospheric ozone has been essential for the evolution of higher forms of life: without it, life on Earth would consist only of UV radiation–tolerant microbes and algae. An oxygenated atmosphere began to develop some 2.5 billion years ago thanks to photosynthesis by oceanic cyanobacteria, and the rising concentrations of tropospheric oxygen eventually led to the accumulation of ozone in the stratosphere, the atmosphere's topmost layer, which extends up to about 50 kilometers above the ground, with highest O_3 concentrations at about 30 kilometers, more than 20 kilometers above the top of Mount Everest.

This ozone shield is transparent to all longer wavelengths of ultraviolet (below the visible range) radiation (UVA between 320 and 400 nm, which is essential for vitamin D formation when absorbed through the skin but which can cause sunburn and cataracts. But stratospheric ozone shields the biosphere from the shortest (the most energetic and DNA-damaging) wavelength of UVB (280–320 nm) by absorbing all wavelengths shorter than 295 nm, thereby allowing the evolution of complex terrestrial and marine life. Marine phytoplankton is especially sensitive to UVB, and depletion of ozone would lead to declines in photosynthetic productivity. UVB radiation also affects the reproductive capacity and larval development of marine animals, while its terrestrial effects would first manifest in cataracts and skin lesions in animals and people, and in reduced crop yields.

In 1975 Rodolphe Zander reported the first clear evidence of CFCs being transported into the stratosphere by identifying the end-product of their photolysis, and before the decade's end two global measurement networks had been put in place. The monitoring showed a steady rise in concentrations, but there still was no proof that the process outlined by Rowland and Molina was actually destroying stratospheric ozone. But precautions began to spread. The global production of CFCs declined from its 1974 peak; in March 1978 the US, Canada, Norway, and Sweden banned the use of nonessential aerosols; and in 1980 the European Community made a commitment to a CFC capacity cap and a 30 percent reduction in aerosol use.

In 1982 and 1983 evaluations by the US National Academy of Sciences forecast that continued use of R-11 and R-12 at 1977 levels would eventually reduce the global ozone level by 2–4 percent rather than by the 10–15 percent as previously forecast, weakening but not ending the push toward a rapid CFC ban. In March 1985 a meeting of forty-three nations resulted in the Vienna Convention for the Protection of the Ozone Layer, which promised to take appropriate control measures to protect the ozone layer and to produce a binding international agreement by 1987—and this need became urgent when on May 1, 1985, *Nature* published a paper that refuted the models predicting that ozone perturbations would remain small for at least the next decade. Its authors, led by Joseph Farman, working with the British Antarctic Survey, reported that the spring concentrations of total O_3 in Antarctica had fallen considerably, a finding that became widely, and quite inaccurately, known as the annual seasonal formation of the Antarctic "ozone hole."

Because the circulation in the lower stratosphere seemed to be unchanged, chemical causes appeared most likely, and the authors suggested that the very low temperatures that prevail from midwinter until after the spring equinox "make the Antarctic stratosphere uniquely sensitive to growth of inorganic chlorine" and that this, combined with the height distribution of UV irradiation specific to the polar stratosphere, could explain the observed O_3 losses. This phenomenon was better elucidated in 1986 when measurements done with balloon sondes made it clear that the chlorine reactions were taking place on the surfaces of polar stratospheric clouds. These findings, combined with the well-known longevity of CFCs—an atmospheric lifetime of between forty-six and sixty-one years for CFC-11 and between ninety-five and 132 years for CFC-12—made it obvious that an effective global intervention was needed.

Decisions made by DuPont, the largest US CFC maker (about half the country's total refrigerant volume and the producer of one quarter of the global output), were critical. Subsequent analyses both praised and criticized the sequence of the company's (sometime inconsistent) CFC-related decisions, but its embrace of an early production ban and its role in supplying, fairly rapidly, commercial alternatives are indisputable. Before the discovery of the Antarctic ozone destruction, DuPont stated its readiness to stop making CFCs if faced with incontrovertible evidence of their harm, and the company's assurances that it could supply alternatives were essential for the acceptance and rapid ratification of an unprecedented global agreement. The industry's promise to provide better substitutes was undoubtedly helped by the fact the new compounds were projected to cost five to ten times the dominant CFC-11 and CFC-12.

Negotiations of a binding international treaty to limit and eventually to ban CFCs were thus concluded, with industry being part of the process, and the original Montreal Protocol, signed in 1987, called for a 50 percent production cut of five of the most commonly marketed compounds. Subsequent amendments required a complete phase-out of all CFCs and of several hydrochlorofluorocarbons (HCFCs). In 1990 the London Amendments to the protocol specified a complete phase-out of the most damaging CFCs by the year 2000 in affluent countries and by 2010 in lower-income nations, and in 1992 the Copenhagen Amendments advanced the year of the first phase-out to 1996.

At the beginning of the twenty-first century, CFC-12 was the only compound whose annual output was still above 100,000 tons, and this production decline was accompanied by plateauing and a slow decline of all atmospheric CFC concentrations. CFCs from old refrigerators that were not properly disposed of (removed and incinerated at high temperature) but simply discarded continued to add to the atmospheric burden long after the ban on production went into effect. Most notably, in China releases of CFC-11 and CFC-12 reached maxima in 2011 and ceased only by 2020. The decline in atmospheric concentrations has been slow but steady. Reconstruction of past levels and monitoring of CFC-11 (since 1977) show the averages for the Northern Hemisphere rising from 0.7 parts per trillion (ppt) in 1950 to 177 ppt in 1980, peaking at 270 ppt in 1994 and then declining to about 225 ppt in 2020.

What effect did the bans and restrictions have on the Antarctic ozone hole? Two measures are relevant: its total area and the intensity of ozone depletion. When the Antarctic ozone measurements began in 1956, concentrations above the continent averaged about 300 Dobson units, and this level prevailed until the mid-1970s. The subsequent decline brought the concentrations to just above 100 Dobson units by 1995, and this was followed by stabilization and a slow recovery (rising minimum concentrations). The UN's 2018 assessment concluded that the continent's ozone layer was on the way to recovery and that pre-1980s levels might return by 2060. But the area of substantial ozone depletion (the "hole's" size) keeps fluctuating. In 2019 it extended over only about 8 million km^2, the smallest on record since its discovery: in 2020 it was three times larger, peaking at about 24 million km^2 in October (for comparison, Antarctica covers 14.2 million km^2), and in 2021 it was even larger, at 24.7 million km^2, the eighth largest since record keeping began in 1979 (fig. 2.7).

And what would have happened in the absence of the Montreal Protocol and its amendments? Of course, the answer is contingent on the rate of CFC production. Because the global output of CFCs was declining for years before 1985, a simulation of a future world with unregulated CFC

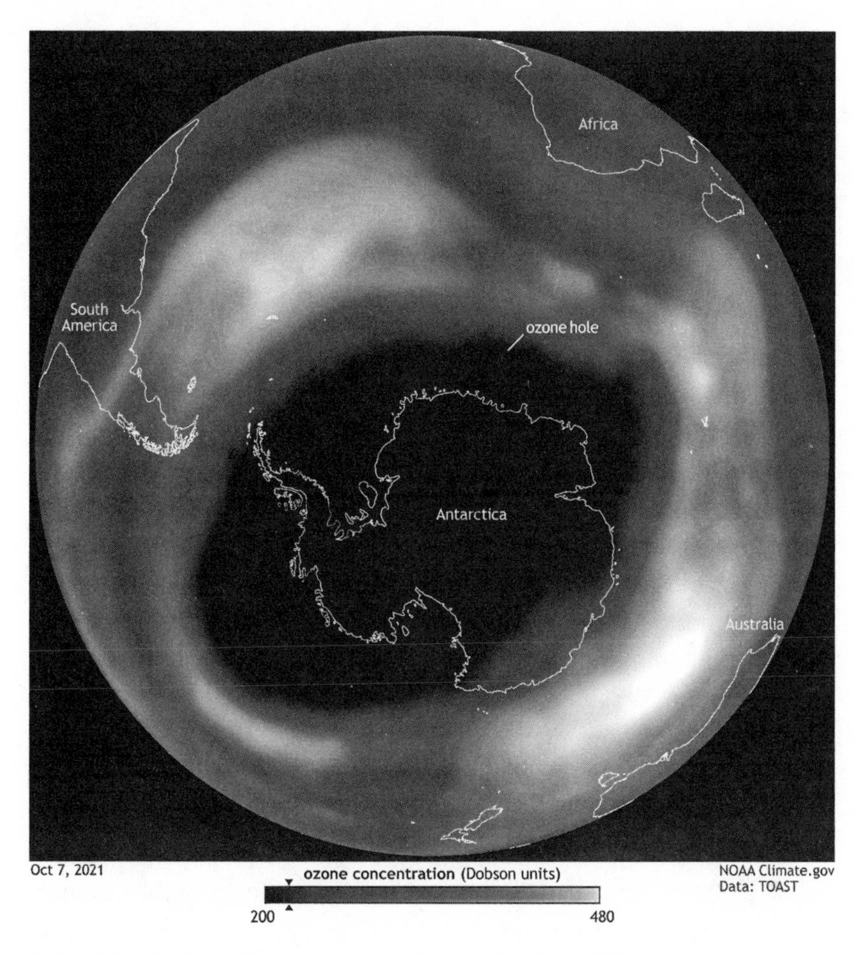

Figure 2.7 Southern Hemisphere ozone levels in October 2021. Low ozone concentrations still persist above Antarctica. *Source:* NOAA Climate.gov.

production growing at 3 percent a year (that is, doubling roughly every twenty-three years), published by Paul A. Newman and co-workers, presents a worst-case scenario. In comparison to 1980, continued growth of output would have destroyed 17 percent of the globally averaged ozone by 2020, and 67 percent by 2065; large ozone depletions would become chronic in polar regions; and by 2060 UV radiation increases would more than double the summer erythemal radiation in the densely populated northern mid-latitudes. Clearly, an impact only a third or a fifth as large would have warranted the measures that were taken.

Perhaps the easiest part of eliminating CFCs were their uses for precision electronics and metal cleaning: that was done by substitution with no-clean fluxes and with water-based solvents. Fulfillment of the Montreal Protocol obligations in the case of such mass-scale uses as refrigeration and air conditioning relied primarily on substituting CFCs by HCFCs, compounds that were known for decades and whose large-scale production and marketing could be realized in a matter of years. Because most of them are removed by chemical reaction while diffusing throughout the troposphere (the atmosphere's lowest layer, extending up to 10–15 kilometers above ground level), their ozone-destroying potential is a small fraction, 1–15 percent, of the most commonly used CFCs.

But HCFCs are not (albeit comparatively less important) only ozone-destroying gases, they are also relatively large contributors to an even more intractable environmental problem, namely, global warming induced by anthropogenic emissions of various "greenhouse gases." When gases are compared on the basis of their global warming potential (GWP) over a period of one hundred years, with CO_2, by far the most abundant gas emitted by human actions set at one, the scores are 28 for methane (from natural gas production and transport, rice fields, and enteric fermentation of ruminants), 265 for nitrous oxide (from fertilizers), 4,160 for the now outlawed CFC-11, and 10,200 for CFC-12—but nearly 2,000 for the most commonly used HCFCs (CH_3CClF_2). The production of these gases was to end by 2040, but in 2007 the high-income signatories of the Montreal Protocol agreed to their phase-out by 2020, and in low-income countries the phase-out began in 2013 and is to end by 2030.

The next available substitutes are hydrofluorocarbons (HFCs). Because they do not contain chlorine, they do not affect stratospheric ozone, and they are not controlled by the Montreal Protocol. Their widespread use is complicated by the fact that they, too, have significant global warming potential, 12,400 for CHF_3 and 1,300 for CH_2FCF_3, the two leading HFCs. In retrospect, it now seems that the search for ideal refrigerants is back to the situation prevailing during the late 1920s when we used whatever worked. The second generation of refrigerants gave us safety and reliability, but the compounds endangered stratospheric ozone. The third generation of refrigerants vastly reduced or entirely removed the ozone problem but contributed to greenhouse gas emissions.

All of this seems to amount to a continuing, and accelerating, sequence of if not failures then repeatedly imperfect solutions. CFCs, ideal synthetic refrigerants that displaced the older natural choices, reigned for nearly half a century; HCFCs in rich countries were dominant for less than forty years; HFCs, devoid of chlorine, have completely eliminated any ozone destruction concerns, but their large-scale use would result in a significant addition to anthropogenic emissions of greenhouse gases, and more so if one considers the enormous future demand for refrigeration and air conditioning in the low-income tropical and subtropical countries of Asia and Africa. By 2020 there were some 1.8 billion air-conditioning units in operation, with more than half of them in just two countries, China and the US. But this is only a fraction of the potential total because among the nearly three billion people living in the world's warmest climates, fewer than 10 percent have air conditioning, compared to 90 percent in the US or Japan.

The need for effective, safe, affordable, and environmentally friendly refrigerants is thus greater than ever. Once again, we need better alternatives, but the enormous post-1930 advances in our chemical understanding do not leave us with any large areas of unexplored options in which we might find new refrigerants that are neither toxic nor flammable nor halogenated (containing chlorine or fluorine) while having the desirable boiling points, low vapor heat capacities, low viscosities, and high thermal conductivities. In the context of the unfolding preoccupation with global warming, choosing low-GWP fluids is imperative, and the possibilities that have been recently studied as potential commercial refrigerants include the old "natural" pre-CFC standbys—carbon dioxide, ammonia and hydrocarbons (ethane, propane, cyclopropane), and dimethyl ether—some fluorinated alkanes (HFCs), and fluorinated alkenes, oxygenates, and nitrogen and sulfur compounds.

If we do not find any promising new candidates, will we be able to accommodate one or two among the old "natural" refrigerants for large-scale uses in homes and cars? Possibly, but not necessarily. But there is one thing we know for sure: unlike the introduction of leaded gasoline, the threat posed by CFCs to stratospheric ozone was a truly unforeseeable failure of innovation. As a result, I find some internet postings regarding Midgley's role in introducing leaded gasoline and CFC refrigerants to be not only grossly exaggerated but plainly inaccurate, nothing but poorly informed historical revisionism that befits the instant expertise characteristic of the medium. "One Man Invented Two of the Deadliest Substances of the 20th Century"; "Thomas J. Midgley is now considered one of history's most dangerous inventors." All that in the century that saw the invention of nuclear weapons, but asking whom to charge with that—Robert Oppenheimer, James Chadwick, Leo Szilard, or a score of other plausible candidates—shows how ridiculous such attributions are.

The indiscriminate bombing of cities—actions that required the invention and major transformation of heavier-than-air flying machines, the extraction and refining of liquid fuels to power them, the development of electronic navigation systems to guide them to their targets, and the use of high-power explosives or incendiary bombs to produce unprecedented degrees of remotely released destruction—killed millions during the twentieth century (the February 1945 raid on Tokyo alone incinerated more than 200,000 people), while no instant (and I suspect very few delayed) fatalities can be contributed to CFCs. And are we to charge Karl Benz, Gottlieb Daimler, and Wilhelm Maybach with some 1.2 million annual deaths from car accidents because they invented the precursors of all modern automobiles?

3

INVENTIONS THAT WERE TO DOMINATE—AND DO NOT

Many fundamental scientific and technical breakthroughs were not recognized as such when they took place. Original publications in specialized journals are read by only a small number of experts, patents are overlooked and forgotten or dismissed as adding nothing new, lost trails of discovery may be reentered only decades later—and only then may they turn into broad thoroughfares leading not only to new industries and new products but also to new modes of social organization and interaction. Perhaps the all-time best example is James Maxwell's formulation and development of the theory of electromagnetic waves, a fundamental advance he was able to accomplish in his writings between 1865 and 1873. Maxwell's ideas provided the foundation for all modern wireless electronics: radios, TVs, mobile phones, the internet, GPS—all these are just higher-order technical elaborations of his fundamental insight.

Among the great twentieth-century advances I cannot think of a better example than the first patent for a solid-state electronic device, granted to the German physicist Julius Edgar Lilienfeld first in Canada in 1925 and then in the US in 1926. For decades the idea of a solid-state amplifier, an invention badly needed to replace large masses of hot glass in vacuum tubes, was attributed to three physicists working at Bell Telephone Laboratories (BTL): in early 1948 John Bardeen and Walter Brattain filed their patent for a germanium point-contact transistor, followed by William Shockley's application for a junction transistor; the three shared the Nobel Prize in Physics in 1956. But BTL eventually admitted (on its memorial website, now defunct) that it had merely reinvented the transistor, and in 1988, four decades after the BTL patents were issued, Bardeen made it clear that "Lilienfeld had the basic concept of controlling the flow of current in a semiconductor to make an amplifying device" but that many years of theory development and advances in material science were needed to turn his idea into a commercial reality.

But ten or twenty years after Lilienfeld's great insight, the only people who might have come across his idea would have been some patent attorneys scouring the archives. In contrast, some scientific ideas and some technical advances have been almost immediately welcome as highly promising and widely seen as significant departures in new directions, as the beginnings of rewarding developments that would solve difficult and persistent challenges and create new markets. The story of penicillin and the subsequent rapid rise of antibiotics—indeed, their excessive overuse!—is an excellent example of these fulfilled expectations. But other innovations have followed disappointing trajectories: they did not develop along anticipated trajectories; their rise came to an abrupt or gradual end, or they declined to insignificance; their ultimate fate has ranged from complete commercial failure to a disappointing stagnation.

Again, as in the first topical chapter, I have selected three prominent examples of these unfulfilled—or at least grossly underfulfilled—early hopes, and I again treat them in chronological order. Airships are lighter-than-air structures that originally, as befits artifacts derived from hot air balloons, had flexible covers. But their later and much larger designs were rigid structures with gas containers arrayed inside. Their development began before the first serious attempts at flight with heavier-than-air airplanes, but both these techniques achieved fundamental advances during the first decade of the twentieth century. By 1909, less than a decade after the first flight of a large rigid airship powered with internal combustion engines, came the world's first airline using a Zeppelin airship; newspapers and magazines were publishing accounts of the impressive flight performance of the new dirigibles and speculating about their coming conquest of intercontinental air travel.

These developments were derailed by World War I, but by 1930 the German Zeppelin was making scheduled flights from Frankfurt to New Jersey, descending toward the Hudson River above Manhattan's skyscrapers. What a demonstration of new flight capabilities, what a promise of future advances! But seven years later the passenger transport in lighter-than-air dirigibles was transformed into nothing but a brief and instantly ended episode in the history of long-distance flight. Compared to this, the unrealized dominance of nuclear fission, an electricity-generating technique that was seen as the ultimate solution to providing the world with clean and affordable electricity, seems to belong to a different category.

After all, nuclear electricity generation was successfully commercialized: reactors are now operating in more than thirty countries on four continents, and in all but two of those countries, the former USSR and Japan, they have accumulated an admirable record of safe and reliable power generation. All that is true, but it is the gap between the promise and the actual achievement that justifies the inclusion of fission in this chapter. In the US, the country that has built more nuclear reactors than any other, the technology that was initially promoted to be so superior that it would be too cheap to meter (this is not an apocryphal reference; Lewis L. Strauss, at that time the chairman of the US Atomic Energy Commission, said so in 1954 at the National Association of Science Writers in New York) became known for its enormous construction cost overruns, and its further development was abandoned largely because of its unprofitability.

Most of the world's countries have not considered any commercial nuclear development—major economies that have stayed away from nuclear power include Australia, Indonesia, Italy, Poland, Thailand, and Vietnam—and fission produced only about 10 percent of world's electricity in 2020 (with national shares ranging from 5 percent in China to 70 percent in France), a small fraction of its contribution anticipated half a century ago. Moreover, the two disasters, at the Chornobyl Nuclear Power Plant in 1985 and at the three Fukushima Daiichi reactors in 2011, reinforced—that is, exaggerated and misinterpreted—fears of nuclear fission: the Japanese plant failure led Germany, the largest EU economy, to terminate its nuclear program, and even the fission's claim to a carbon-free electricity generation has not sufficed to make it a key ingredient of the recent global quest for decarbonized economies.

My last example of unfulfilled promise is the quest for supersonic flight, that is, for transportation at speeds surpassing (when measured at sea level and at 20°C) 1,235 kilometers per hour (km/h). This was a purely science fiction speculation at the time the first airships and, soon afterward, the first airplanes began transporting passengers before World

War I. Even speeds only half that rate were impossible as long as the reciprocating (piston) gasoline-fueled engines were the only available prime movers, but much higher speeds became achievable with jet engines (gas turbines), the new mode of internal combustion that dispensed with cylinders, pistons, and valves, relying instead on continuous combustion to generate powerful propulsion. Expectedly, such speeds were first achieved by military aircraft during the late 1940s.

Once the jetliners entered scheduled commercial service during the 1950s, many engineers and some governments believed that the obvious next step was to increase their cruising speeds (at the time they were flying at about 85 percent of the speed of sound) to supersonic rates. This would cut the tedium of intercontinental travel by half or more, a performance that has obvious commercial appeal but that faces many technical and environmental barriers. Flight aficionados know how this decades-long quest eventually failed, and in the last section of this chapter I recount this high-tech saga and note some of the recent efforts aimed at resurrecting supersonic aviation, this time starting with smaller business jetliners.

AIRSHIPS

The early twenty-first century's reigning perspective on flight is dominated by the successful evolution of many heavier-than-air plane designs that eventually created a massive global system that in 2019 (pre-COVID-19) handled nearly 4.5 billion passengers on more than 38 million flights and totaled about 8.7 trillion revenue passenger-kilometers. In comparison to large (seating hundreds of passengers) yet sleek-looking modern jetliners, lighter-than-air (LTA) flying machines appear clumsy, outmoded, painfully slow, hopelessly inefficient, and incorrigibly weather dependent, and hence unfit for any mainstream use in modern aviation. But that most definitely was not the consensus opinion, expert or public, during the first four decades of the twentieth century, or more precisely until 1937, the year when the *Hindenburg*, trying to land in Lakehurst, New Jersey, after yet another uneventful transatlantic crossing, burst into flames, resulting in what is to this day one of the best-known, instantly documented catastrophes.

The history of LTA flight began with adventures in ballooning. On September 19, 1783, Joseph-Michel and Jacques Étienne Montgolfier filled their balloon (made of cotton canvas and glued-on paper) with hot air, loaded the wicker basket with three small animals, a sheep, a duck, and a cockerel, and let it rise (tethered) in front of the king and a curious crowd. Like all (hot air or light gas) balloons, their small fabric design was a passive object, either carried by prevailing wind or immobilized by calm. The history of steerable LTA flying machines whose direction and speed could be controlled by a prime mover began, abortively, less than a year after the Montgolfiers' demonstration, when brothers Anne-Jean and Nicolas-Louis Robert attempted to propel a small elongated hydrogen-filled balloon with oars. During the same year, 1784, Jean Baptiste Marie Charles Meusnier designed a much larger elliptical airship that was to be powered by hand-cranked propellers, another fantastically impractical idea.

Nearly seven decades passed before Jules Henri Giffard launched the first real airship, on September 24, 1852. His *dirigible* had a steerable cigar-shaped nonrigid envelope that was 44 meters long, its volume (filled with coal gas) was 3,200 cubic meters, and it was powered by a 2.3-kilowatt steam engine that weighed 113 kilograms and required a 45.4-kilogram boiler turning a three-bladed propeller. This setup, heavy and unwieldy, was still too weak to fly into the prevailing winds, and the dirigible could circle only slowly, managing no more than 10 km/h and covering just twenty-seven kilometers between Paris and Élancourt.

More than thirty years elapsed before a duo of French officers, Charles Renard and Arthur Constantine Krebs, performed the first completely controlled-powered round trip with a dirigible, *La France*, on August 9, 1884. Their elongated balloon had a volume of nearly 1,900 cubic meters and the airship was propelled by a battery-powered electric motor turning a wooden propeller seven meters in diameter. After covering eight kilometers in twenty-three minutes, they landed on the parade ground from which they took off. Additional flights followed in 1884 and 1885. The first small airship powered by an internal combustion engine was demonstrated by Friedrich Wölfert in Berlin in 1897, and the real breakthrough in powered airship flight began in 1899 when Ferdinand, Graf von Zeppelin, at that time a retired (dismissed) German army general,

Figure 3.1 Ferdinand Adolf August Heinrich, Graf von Zeppelin (1838–1917), an indefatigable pioneer of airships for long-distance passenger transport.

turned to building his rigid designs using aluminum, an impermeable cover, and (suspended or directly attached) gondolas (fig. 3.1).

Zeppelin's interest in LTA flight dated back to his brief visit to the US, first as an observer of the Civil War with the Union forces, then as a visitor to the country's expanding western frontier: in Minneapolis he ascended in a balloon inflated with coal gas (previously used for observations by the Union Army). His diaries from a decade later describe the basics of his signature airship design, a rigid airship made of rings and longitudinal girders filled with individual gas cells, but only in 1890, after his forced retirement from the army when he was fifty-two, did he turn to designing and building LTA machines. He piloted the first flight of the *Luftschiff Zeppelin* 1 (LZ-1) on July 2, 1900.

Larger airships followed, some acquired by the army, some, unmoored, destroyed by wind gusts and fire. Deutsche Luftschiffahrts-Aktiengesellschaft (DELAG), the world's first passenger airline, was set up in November 1909, and before the beginning of World War I more than 1,500 people had flown on 218 scheduled domestic flights. LZ-13, the *Hansa*, launched in July 1912 and set new commercial records, flying nearly 45,000 kilometers over the course of 399 flights and visiting Denmark and Sweden—at a time when airplanes were still small wood-and-canvas affairs. Airships seemed to be the next big thing in long-distance transportation.

In 1912 Thomas Rutherford MacMechen and Carl Dienstbach wrote about the "greyhounds of the air," noting that "for yet a little time" the great ocean liners would continue their

> boastful voyages, and perhaps for another decade nations will waste their treasures upon floating fortresses. But the end is near. Tomorrow those who wish to hasten across the Atlantic will take an airship. For them the crossing will be one of hours.

And not only that, Great Britain, that mistress of the seas able to enjoy "a dream of complacent confidence," would have to face, as the next war was to demonstrate, the threat from the air, but an airship as a weapon might be so terrible "that it may be a powerful factor in furthering world peace." For the writers, these were not theoretical musings because demonstrations and proofs (first test flights of the Zeppelin over the ocean) were ready: much larger airships would double or triple the speeds of the fastest ocean liners, "nothing but its size limits a distance a Zeppelin can cover, and the limit of practical size is nowhere in sight"—and "venturesome persons are indeed planning flights across the Atlantic for the near future."

Zeppelin's designs benefited from the availability of the new lightweight and powerful internal combustion engines, as well as from the new possibilities of radio communication. In 1908, Wilhelm Maybach, the cocreator of the world's first automobile and the designer of the Mercedes 35 (generally considered to be the first truly modern car prototype), and his son Karl began to build the engines for Zeppelin's airships. World

War I interrupted further development of passenger airships, but the German military became a large-scale customer: it acquired nearly 140 airships to be used for aerial reconnaissance and as long-distance bombers.

More than one hundred of them were Zeppelins (the LZ-26 model was first launched in 1914 with a volume of 25,000 cubic meters, a length of 161 meters, a payload of three tons, and a range of 3,300 kilometers); the rest were built by Schütte-Lanz, the country's second airship builder, which the German government forced to cooperate with Luftschiffbau Zeppelin. The war's first airship raid was on Liège on August 6, 1914, and it was the LZ-17, which, after carrying nearly 10,000 prewar passengers and flying nearly 40,000 kilometers, was converted to a bomber; a second attack, on Antwerp on August 25, 1914, followed. Bombings of France and England ensued, with air raids (foreshadowing the Battle of Britain) causing thousands of civilian casualties and substantial material damage. England was attacked for the first time on the night of January 19–20, 1915.

Initially, the country was defenseless, but by 1916 a combination of artillery, searchlights, fighter aircraft, and the ability to intercept German radio messages had changed that, and in 1917, 77 of the 115 German airships sent on bombing raids were either shot down or fully disabled. An unexpected attempt took place in November 1917 when an LZ-104 (the *Afrika-Schiff*) embarked on an unprecedented long-range logistical airlift to resupply German colonial troops in East Africa. The 226.5-meters-long ship, powered by five 180 kW Maybach engines, started in Bulgaria, crossed the Mediterranean, and made it as far as central Sudan (west of Khartoum) before getting recalled and reaching Bulgaria after covering 6,800 kilometers in ninety-five hours. Count Zeppelin died before this aborted attempt (on March 8, 1917), and after the war Hugo Eckener, originally a psychologist but from 1911 a certified airship pilot, took over the company, whose future was uncertain because the peace treaty forbade any further construction of German airships.

The first remarkable postwar accomplishment came in July 1919 when the British dirigible R-34 made a round trip between Scotland and New York's Long Island, the first LTA machine to cross the Atlantic, just a month after John Alcock and Arthur Brown flew their modified Vicker Vimy bomber from St. John's, Newfoundland, to Clifden in Ireland. But there were no further notable British developments of LTA flights, and

Figure 3.2 *Graf Zeppelin* airship above the German Reichstag on October 1, 1928. *Source:* Bundesarchiv photo 102–06617.

the peace treaty restricted Germany's airship construction. In October 1924 an LZ-126 (renamed the *Los Angeles*) was delivered to the US as a part of German war reparations and served the US Navy until 1940. After the treaty restrictions were eased in 1925, Hugo Eckener, chairman of Luftschiffbau Zeppelin, mobilized public and government support to build a new passenger airship intended as a prototype of even larger and faster commercial designs. The *Graf Zeppelin* flew for the first time in September 1928, and during its fewer than nine years in service it accomplished many aviation firsts (fig. 3.2).

In 1929 the airship toured southern Europe, the Middle East, and Africa, but its greatest accomplishment was the circumnavigation of Earth, half-financed by William Randolph Hearst. The airship flew eastward from Lakehurst, New Jersey, to Friedrichshafen, then on to Tokyo and Los Angeles, returning to New Jersey three weeks after its departure. The following year it flew from Germany to Brazil and the US; in 1931 it was on an Arctic expedition, and that same year it began regular passenger and mail service between Germany and Brazil. At that time Zeppelin had no long-distance competitor among existing, heavier-than-air airplanes. In 1931 Boeing's Monomail (which in a later version carried six passengers and mail) had a range of just 925 kilometers.

The only way to do trans- or intercontinental travel was in tedious stages: three stops and more than fifteen hours were needed to make it from New York to Los Angeles, and when British Imperial Airways began to operate the London-Singapore link in 1934, its planes needed eight days and twenty-two layovers, including stops in Athens, Cairo, Baghdad, Basra, Sharjah, Jodhpur, Calcutta, and Rangoon. And while the Douglas DC-3—introduced in 1935 and destined to become the most common and most durable piston-powered airplane in history—was about twice as fast as the Zeppelin (240 km/h), it had a maximum range of about 2,500 kilometers, just a quarter of the Zeppelin's reach. And the cramped interiors of the first small all-metal-fuselage airplanes of the early 1930s provided no comparison to the overall roominess, designed public lounges, and dining room of a large airship.

By the time it was grounded, in June 1937, the *Graf Zeppelin* had flown 1.7 million kilometers, carried more than 13,000 passengers, completed 144 intercontinental trips, and spent 717 days—nearly two years—aloft, all, despite some in-flight mishaps, without an injury to its crew and passengers. Unlike all previous Zeppelin designs, the next airship was to be filled with inert helium rather than with inflammable hydrogen. But the helium supply (with the gas extracted at hydrocarbon fields) remained controlled by the US, and the Helium Control Act of 1927 expressly forbid the element's export. That decision was, not surprisingly, maintained after the Nazis came to power in Germany; they eventually ousted the anti-Nazi Hugo Eckener and put swastikas on the airship's fins. The LZ-129, named *Hindenburg* after Germany's World War I field marshal and president (1925–1934), was launched on March 4, 1936. The *Hindenburg* was the world's largest airship at 245 meters long and just over 41 meters in diameter, with a volume of 200,000 cubic meters, powered by four Daimler-Benz diesel engines (890 kW each) and cruising at 122 km/h.

The airship was used first for domestic test flights and Nazi propaganda flights. Subsequently it made seventeen intercontinental trips, seven to Brazil and ten to the US: few propaganda images could equal those of the *Hindenburg*, marked with a swastika, descending over Manhattan toward its New Jersey landing. Passenger capacity was increased from fifty to seventy, the interior design of the airship's public spaces and viewing galleries was praised as much as its smooth take-offs and stable flights, but although the 1936 intercontinental flights were scheduled, all of them were still within the test or demonstration stage. The first US-bound commercial flight of 1937 departed Frankfurt on May 3 and its landing at Lakehurst on May 6 ended catastrophically, with thirty-five of the ninety-seven people on board dead.

There had been previous airship disasters with significant human casualties, but except for the British R101, brought down by a storm during its first long-range test flight over France in 1929, killing forty-eight people, all involved military crafts: the Royal Navy's R38 in 1921 (forty-four dead), the US Army's *Roma* in 1922 (thirty-four dead), the French *Dixmude,* a former Zeppelin, in 1923 (fifty-two dead), and the US helium-filled *Akron* in 1933 (seventy-three dead). But the *Hindenburg* was different: it was the first commercial airship destroyed by an explosion and fire, documented as it occurred, and the first of Germany's long line of Zeppelins dedicated to passenger transport.

The last minutes of this catastrophe, including the initial fire, explosion, and the crash, were filmed by at least five different news services, Pathé News, Paramount News, Movietone News, Universal Newsreel, and News of the Day, making it "the first media event of the twentieth century." Subsequent analyses explained how a cascade of unlikely events produced an unpredictable catastrophe: of course, hydrogen-filled airships had always posed risks, as does any form of transportation, but the previous safety record of German rigid dirigibles and their numerous uneventful Lakehurst landings made this specific event unpredictable. Arguments about the inevitability or prevention of the catastrophe became instantly irrelevant: the "made-for-movies" disaster was too spectacular to continue the flights. The short era of German hydrogen-filled passenger airships ended suddenly: the next ship in line, the LZ-130, was completed in 1938 but made only some military reconnaissance flights before it was deactivated.

World War II saw the return of military airships. Barrage balloons were used in the US, Europe, and Japan, but the US was the only major power that used a large number of airships. Goodyear's K-series airships were dominant. They were neither large, having a volume of 12,000 cubic meters and a length of 76 meters, nor very fast, with a top speed of 80 km/h, but, powered by two 317 kW engines, could stay aloft for up to sixty hours. The US Navy deployed them for minesweeping, search and rescue, reconnaissance, scouting, antisubmarine patrols, and, perhaps most notably, escorting ship convoys. In total, airships patrolled nearly eight million square kilometers of the Atlantic and Pacific Oceans and the Mediterranean Sea, and only one was shot down by a German submarine.

Military LTA designs did not disappear entirely after the war. Between 1952 and 1962 the US Navy had a secret program using ZPG-class airships to fill gaps in North America's early radar warning system: they could stay on station for more than two hundred hours and conduct long unrefueled patrols. By the 1960s these roles were being filled by new superior reconnaissance airplanes and, in a completely safe manner, by satellites—but the airship lobby never gives up. Northrop Grumman built a prototype of a surveillance airship in 2012, but the contract was canceled the next year. In 2015 Raytheon, another military contractor, lost its prototype JLENS (Joint Land Attack Cruise Missile Defense Elevated Netted Sensor System) spy airship after it broke its tether and drifted above Pennsylvania, effectively ending the development contract that began in 1998. But I have no doubt that the well-established US military-industrial complex will conjure up new needs for new aerostats or dirigibles; that stream of funding will keep on flowing.

In contrast, any realistic prospects for commercial airships on intercontinental routes ended even before World War II, and they did so not because of the *Hindenburg* catastrophe but because of advances in airplane propulsion. Until the mid-1930s no airplane could compete with airships in combining passenger capacity (up to seventy people) with maximum range (up to 10,000 kilometers): LTA airships had a clear advantage in pricey but reliable and safe intercontinental passenger transport. But even if the *Hindenburg* had continued to fly with a perfect safety record, it was an anachronism by the time it was launched. In July 1936, just four months after the *Hindenburg* was launched, PanAm airlines signed a deal with Boeing to get the first six of the company's new B-314 Clippers, large hydroplanes that could carry up to sixty-eight passengers and cruise at just over 300 km/h. The plane began its scheduled transpacific San

Francisco to Hong Kong service in February 193 and also made its first prewar trip to England.

Moreover, even during World War II it was clear that the dominance of aviation piston engines would end soon as the first newly deployed military jet engines (gas turbines) would reach the commercial market and enable long-distance transportation at speeds approaching the speed of sound. Indeed, in 1952 the British Comet, the world's first, and ill-fated, commercial jetliner, had a cruising speed of nearly 740 km/h, and in 1958 the very successful Boeing 707, the beginning of the longest series of jetliners, worked best at 897 km/h, very close to the latest Boeing 787 at 913 km/h. This, of course, was nearly an order of magnitude faster than the Zeppelin's typical speed, and while the *Hindenburg*'s fastest crossing times between Frankfurt and New Jersey were nearly fifty-three hours westward and forty-three hours eastward, today's scheduled flying times by Boeings or Airbuses are, respectively, eight hours thirty-five minutes and seven hours twenty minutes, and under far more controllable circumstances.

Airships for passenger transport may be gone, but dreams of airships in other roles—above all as cargo carriers and platforms for scientific studies and for military reconnaissance—keep recurring and collapsing. The most spectacular of these failures was Germany's giant CargoLifter. The company, established in 1996, issued stock and received plenty of federal funding; its ultimate goal was an airship of 550,000 cubic meters (nearly three times the *Hindenburg*'s volume) able to lift 160 tons. The monster was never built (though a giant hangar for it was), the company went bankrupt in 2002, and the hangar, the largest freestanding structure ever built, is now a tropical water park—but CargoLifter's website is still on the web, promising future LTA wonders.

And it is not alone. Recent promoters of airships always note how the advances in materials, propulsion, and electronic controls could combine to produce a highly functional, very reliable, flexible, and economically acceptable (and also more sustainable) LTA cargo lift solution. The US military has never let go of the idea, with reexamination occurring every few years, spurred by experiences in recent American wars and in search of airships to use both for cargo lift and as high-altitude surveillance and communication platforms operating at altitudes between 18 and 24 kilometers. Cargo airships are seen to increase the flexibility, availability, and service life of the strategic jet airlifter fleet, while high-altitude airships could provide coverage that is now available only with unmanned aerial vehicles, and could do so over much longer periods.

Commercial applications do not lack their promoters, some even claiming that an emerging international competition will lead to the airships' return because, as the persistent sales pitch goes, they are relatively cheap, can carry substantial loads, and, perhaps the most appealing consideration in the early 2020s, their operation produces only a small fraction of greenhouse gas emissions compared to other modes of airborne transport. All of these rigid airship designs have metal frames that support the engines, control surfaces, and cargo hold, with the lift provided by a series of nonpressurized helium-filled cells.

Using airships for cargo lift in the Arctic is an old idea whose realization is now seen as a valuable addition to the region's development (that is, exploiting its stranded resources) in a warming world. These changes give easier access by shipping, but as oil spills are notoriously difficult to contain and clean up in cold waters, while land access, currently over frozen winter routes, may become more restricted as a result of the melting permafrost, making the seasonal resupply of remote communities by trucks even more perilous. According to Barry Prentice, cargo airships are being promoted as "the only conceivable means of transport that can carry large bulky loads over long distances and operate in areas devoid of established infrastructure." Another potential use is to fly fresh fruit and flowers from their subtropical and tropical production sites to major markets of the Northern Hemisphere, but airlifting Hawaiian pineapples to California would have been unprofitable even in 2004, when the proposal was made, because cultivation of this fruit has been in steep decline since the 1970s and the islands are now only minor pineapple producers.

Still, in 2020 it was not a sci-fi-inclined tech website but *Foreign Policy*, a bimonthly journal of the American establishment, that headlined a story "The Age of the Airship May Be Dawning Again," and went on to recount how several companies were trying to bring back "spectacular dirigibles." Another 2020 report, this time in the *Robb Report*, is headlined "These New Luxury Blimps Hope to Become the Superyachts of the Skies." The Luftschifftechnik Zeppelin was revived in 1993 at its original location on Lake Constance. In September 1997, when the first Zeppelin NT (New Technology) took off, the company claimed that "the myth of the Zeppelin was successfully reborn."

It got the myth part right: despite the new designs and new materials, including a rigid triangular structure, helium, a tear-resistant envelope, swiveling propellers, and modern "fly-by-wire" avionics, the company does not have a long line-up of new orders. The same is true of France's Flying Whales, established in 2012. The company has as an enticing website that opens with an animated image of a humpback whale gracefully rising through trees above the forest canopy and into the blue atmosphere, and proceeds with color images illustrating how its 200-meter-long rigid airships with a sixty-ton payload can be used for logging remote forests and transporting wind turbine parts and high-voltage towers to inaccessible places. The US government funded the *Dragon Dream* airship, which had a flattened elliptical cross section; the prototype was heavily damaged after some tethered trials in 2013 (fig. 3.3).

Figure 3.3 One of several failed modern airship revivals: the *Dragon Dream* demonstration design model near its hangar. Image by Aeros (now defunct). *Source:* Parkhannah, https://commons.wikimedia.org/wiki/File:Dragon_Dream.jpg.

Sweden's Ocean Sky Cruises, not missing a contemporary beat, has been promising sustainable "decarbonized aviation" with North Pole excursion flights for years, with the first luxury flight season (eight double cabins "fully equipped with large panoramic windows, a private bathroom, and a small wardrobe") now listed for 2024–2025. In the US, the CEO of Worldwide Aeros Corporation, established in 1993, claimed in 2016 that the company would have a global fleet of Aeroscraft airships operating by 2023. And in 2006 Lockheed Martin, a leading military contractor, tested the P-791, an experimental tri-hull hybrid airship with payload supported by both buoyant and aerodynamic lift, designed for cargo transport to otherwise inaccessible areas.

There is also Lighter Than Air Research, an aerospace R&D company funded by Google cofounder Sergei Brin, which believes its airships will "complement—and even speed up—humanitarian disaster response and relief efforts," especially in those remote areas that cannot be easily accessed by plane and boat, and that ultimately, its "family of aircraft with zero emissions" will be shipping goods and moving people with a reduced global carbon footprint. And to top it all is the Russian Airship Initiative Design Bureau Aerosmena (AIDBA), which announced the projected launch of a giant saucer-shaped airship in 2024: it is to have a maximum payload of 660 tons, a diameter of more than 240 meters, and be powered by turboprops turning helicopter-like rotors. But perhaps the most notable 2022 addition is the order for 10 helium-filled electricity-powered *Airlander* (Rethink the Sky) airships able to carry 100 passengers by Valencia-based Air Nostrum company that plans to operate them on short domestic routes starting in 2026.

All of these claims and plans have one thing in common: they pay little attention either to what any rapid expansion of LTA fleets would do to the supply of helium or what the actual revenue-earning time aloft might be. In the US, recent domestic helium consumption has been about 40 million cubic meters a year, with major uses in magnetic resonance imaging (30 percent), lifting gas (17 percent), and analytical and laboratory applications (14 percent). If all of this annual use went into airships, it would be enough for about two hundred large (Zeppelin-like) structures. The global resources of helium are estimated at about 50 billion cubic meters, with 40 percent in the US, 20 percent in Qatar, and most of the rest in other natural gas–rich countries such as Algeria, Russia, and Canada. As for flying frequency, modern jetliners spend about 3,000 hours per year (about 34 percent) in flight; would large fleets of proposed airships match that? And sci-fi-inclined stories even mention vacuum airships, filled with nothing but with their shells able to withstand atmospheric pressure.

The skies may not be teeming with all these promised neo-Zeppelins, but despite the fundamental challenges posed by volume, gas containment, and flight control, the lure of LTA craft will probably never disappear. In all heavier-than-air machines lift comes only externally, while LTA airships combine internal lift from the natural buoyancy of light gases with the external lift provided by their engines. This makes them more difficult to control in flight and more challenging to design. In any case, most of the LTA designs did not become, as their creators often intended, the prototypes of successful commercial series: airships were expected to dominate, but instead they became asterisks in flying history, marginal accessories in the enormously expanded global world of flight. It is a safe bet that this reality will not fundamentally change anytime soon.

NUCLEAR FISSION

The controlled release of energy from the nuclear fission of uranium went from theoretical concepts to the first commercial electricity generation in exactly sixty years, a remarkably brief period of time if one considers the inherent complexities of the challenge. The first theoretical foundations were laid in the spring of 1896 by Henri Becquerel's discovery of uranium's radioactivity, eleven years later by Albert Einstein's famous conclusion that "an inertial mass is equivalent with an energy content μc^2," and between 1911 and 1913 by Ernest Rutherford's model of atomic nuclei and Niels Bohr's structure of the nucleus surrounded by orbiting electrons.

The next series of fundamental advances came in the 1930s: in 1931 the first splitting of a light element, lithium (using high-voltage electricity to accelerate hydrogen protons), into two helium atoms, and in 1932 James Chadwick concluded that the only way to explain some experiments done in Germany and France was to assume the existence of particles of mass 1 and charge 0; thus were neutrons born. In London, a little more than six months after Chadwick published his neutron discovery, Leo Szilard, an exiled Hungarian physicist and Einstein's student collaborator, had his epochal epiphany as he waited for a green light on Southampton Row when he realized that "if we could find an element which is split by neutrons and which would emit *two* neutrons when it absorbed *one* neutron, such an element, if assembled in sufficiently large mass, could sustain a nuclear reaction."

He did not know how to find that element, but on March 12, 1934, he applied for a British patent that listed beryllium as the most likely candidate for the splitting and that also named, correctly, uranium and thorium as other candidates. Szilard's patent application was kept secret, but he was not the only scientist thinking about neutrons. In Germany, Otto Hahn's and Fritz Strassman's irradiation of uranium by neutrons produced new isotopes, and in February 1939 Hahn's longstanding collaborator, Lise Meitner, by that time living in exile in Sweden, and her nephew, Otto Frisch, correctly interpreted the result as nuclear fission. The atom was split, and the consequences became obvious to all well-informed physicists: on the one hand, the unprecedented destructive power of weapons, on the other, the possibility of a new way to generate electricity.

The outcome is well known. World War II began less than seven months after Meitner's confirmation, and although all major belligerents (the US, USSR, Germany, Japan) pursued the development of nuclear bombs, only the US, with its unprecedented Manhattan Project, aided by Britain, and with the participation of many exiled European physicists, succeeded before the war's end and dropped the first two available weapons on Hiroshima and Nagasaki. A key part of the expensive (and still ongoing) arms race that followed the end of World War II was to put nuclear warheads on virtually invulnerable submarines, and the only way to make these submarines travel far and stay submerged for long periods of time was to propel them by the power of controlled nuclear fission. Again, the US was first in this race as it launched, under the aggressive leadership of Hyman Rickover, its first nuclear submarine in 1954.

Even during the war, some Manhattan Project physicists considered the possibility of using nuclear reactors for electricity generation and concluded that it would be highly uneconomical. That sentiment prevailed even after the establishment of the US Atomic Energy Commission (AEC), and during the late 1940s and the early 1950s it was shared by America's leading electricity-generating companies. Besides the prohibitive cost, there were no compelling resource or environmental reasons to develop nuclear electricity. The US led the world in electricity generation, but the country was also the world's largest producer of fossil fuels, and new large generating plants fueled by coal, oil, and natural gas not only were able to cover the rising needs but were doing so at a lower cost to consumers, leading to mass-scale acquisitions of new household and industrial converters. In the early 1950s, less than a decade after the end of World War II, no country had any strong antipollution policies or movements, and global warming associated with the combustion of fossil fuels and the resulting need for zero carbon energy were to remain outside political and economic realms for more than another three decades.

Why, then, did the US decide to build its first nuclear power generating station? During the late 1940s, David E. Lilienthal, the AEC's first chairman, began to talk about easing the sense of guilt over Hiroshima; he believed that the peaceful development of fission was essential to provide the longed-for psychological relief as well as hope and opportunities for new technical advances. But this sentiment was only one of the factors that changed the nuclear outlook. The USSR tested its first bomb in 1949, and Lilienthal feared that the Russians would beat America "at developing the peaceful side of the atom." The UK committed to a fairly bold nuclear power generation program based on a newly developed domestic reactor design, and in early 1953 President Eisenhower agreed with Secretary of State John Foster Dulles that "it would look very bad if the United States lagged behind" in commercializing nuclear electricity generation. Politics, not economics, dictated the country's development of nuclear electricity generation.

Given these realities, and despite professed doubts by leading nuclear scientists and the country's utilities, the US had to join the quest, especially as such developments could be used as a Cold War tool to influence neutral nations. President Eisenhower spelled this out clearly in his Atoms for Peace speech:

To hasten the day when fear of the atom will begin to disappear from the minds of people and the governments of the East and West . . . to apply atomic energy to the needs of agriculture, medicine and other peaceful activities. A special purpose would be to provide abundant electrical energy in the power-starved areas of the world.

The atomic power race was on, and for the US the most expeditious route was, literally, to beach the small pressurized water-cooled reactor developed by the navy, under the leadership of Hyman Rickover, for nuclear submarine propulsion. *Nautilus,* the US Navy's first nuclear-powered submarine, was built between June 1952 and January 1955 (fig. 3.4). Submarine thermal reactors, built by Westinghouse, used water (pressurized to as much as 16 MPa) in a closed loop to cool the core (packed with fissionable isotope of uranium emplaced in zirconium steel tubes); heated water transferred its energy to a secondary circuit to generate steam for turbines. The same reactor design was used for America's first commercial nuclear project. Pennsylvania's Duquesne Light agreed to share a smaller part of the project's cost, and Shippingport, America's first fission power plant—a symbol of "peaceful and purposeful America"

Figure 3.4 The *Nautilus* in New York Harbor in 1958. *Source:* US Navy photo courtesy of the US Navy Arctic Submarine Laboratory.

Figure 3.5 Emplacing the reactor containment vessel at the Shippingport, Pennsylvania, plant in 1956. *Source:* Library of Congress photo.

that was to generate respect for the country in a nonthreatening way—began to generate electricity on December 18, 1957, nearly six months after the Soviet Obninsk plant and almost fifteen months later than the British Calder Hall plant (fig. 3.5).

This purely political decision tied the country's future nuclear development to a reactor that none of the Manhattan Project physicists and no utility experts considered to be the best option, and this, together with the predicted high generation costs, left the utilities as uninterested in nuclear power during the late 1950s as they were a decade before. Another political decision followed: Congress intervened, and in 1957 it passed the Price-Anderson Act, which made it far less risky to invest in nuclear power generation by reducing private liability and guaranteeing unprecedented public indemnification in the event of a catastrophic accident that released ionizing radiation.

There was no immediate surge of commercial interest, and the next stage began only in December 1963 when an analysis by the Jersey Central Power and Light Company concluded that its planned nuclear power plant at Oyster Creek would generate electricity less expensively than a coal-fired plant. The construction permit came a year later, and in November 1965 a massive power blackout in the US Northeast provided a wider incentive to invest in the new form of generation. New utility orders rose to twenty reactors in 1966 and thirty in 1967 before dipping below ten reactors in 1969, for a total of eighty-three new reactors between 1965 and 1969. But the next year yet another turn of affairs favored nuclear generation: Congress passed the Clean Air Act of 1970, aimed at limiting emissions from mobile as well as stationary industrial sources by imposing the National Ambient Air Quality Standards and New Source Performance Standards. Large coal-fired electricity-generating plants were the nation's leading emitters of particulate matter and sulfur and nitrogen oxides (contributors to acid rain), while nuclear power generation emitted none of these air pollutants.

But fission received by far its biggest boost in 1973 thanks to yet another unexpected series of events. In 1973 the Organization of Petroleum Exporting Countries (OPEC), established to secure better prices for its globally needed fuel, took advantage of weakening US oil extraction (though the US remained the world's largest producer until 1977) by quintupling its posted crude oil price and even temporarily blockading oil exports to the US. The ensuing energy crisis in the US ended the long period of rapid post–World War II economic growth and complicated prospects for establishing a reliable energy supply. Domestic nuclear electricity generation, independent of unpredictable imports, looked very appealing: in 1973 US utilities ordered forty-two new nuclear reactors. Moreover, there was a growing consensus that this rapidly unfolding first nuclear era would soon be followed by a second era of far more effective fast breeder reactors.

They, unlike fission reactors (whether water- or gas-cooled), which operate by splitting abundant isotope ^{238}U, or slightly enriched ^{235}U (from its natural presence of 0.7 percent to no more than 3–5 percent), would use a highly enriched isotope of uranium-235 (15–30 percent) as the source of fast neutrons to convert abundant but non-fissionable isotope ^{238}U, placed in a blanket surrounding the reactor core, to fissile plutonium (^{239}Pu). Liquid sodium would transfer the generated heat and a breeder reactor would eventually produce at least 20 percent more fissionable fuel than it consumed. Szilard had envisioned breeder reactors already in 1943, and in 1945 Alvin Weinberg and Manhattan Project physicist Harry Soodak conceptualized their design.

After World War II, small breeders were tested in both the US and the USSR, and by the early 1970s experimental liquid metal fast breeder reactors (LMFBR, the metal being liquid sodium used for cooling) were operating not only in the US and USSR but also in the UK, France, Germany, Italy, and Japan. In 1973 Alvin Weinberg did not have "much doubt that a nuclear breeder will be successful" and that it was "rather likely that breeders will be man's ultimate energy source." This appeared to be an unexceptional conclusion as there was a general scientific consensus about the desirability of this technique and its eventual success, a conviction shared by industry leaders. During the late 1960s and early 1970s the AEC projected one thousand reactors operating in the United States by the year 2000, and in 1974 General Electric predicted that breeders would be commercially introduced by 1982, all fossil-fueled energy generation would be gone by 1990, and by the century's end all but a tiny fraction of the electricity used in the US would come from breeder reactors.

The reality proved quite different, and two French terms seem quite apposite here: the *dénouement* was a *débâcle*. Its major contributory causes were the unanticipated sudden end to the decadal doubling of electricity demand, excessive regulatory measures imposed on new plant construction, the ensuing mass cancellations of pressurized water reactor orders, failure to transform breeders from physicists' dreams into even a semi viable engineering reality, and rekindled public distrust of fission used for electricity generation as a result of catastrophic accidents. I will explain briefly each of these factors, but I do not attempt to assign proportional blame (I am not sure that is even possible).

But it is clear that no single factor was more important than the rapid retreat of electricity demand. Annual growth around 7 percent, resulting in a doubling of demand every decade, was, with the exception of the slowdown during the crisis decade of the 1930s, the norm since the end of World War I: during the 1920s electricity generation had almost exactly doubled, during the 1940s it grew nearly 2.2 times, between 1950 and 1960 it rose nearly 2.3 times, and between 1960 and 1970 it once again and almost exactly doubled; but during the 1970s it grew by a bit less than 50 percent, followed by about 33 percent growth during the 1980s, 25 percent growth during the 1990s, less than 9 percent gain during the first decade of the twenty-first century, and no growth at all between 2010 and 2019 (a comparison that includes COVID-19-affected 2020 results in a 3 percent decline).

By the early 1970s the largest new, typically multi-unit thermal stations had capacities in excess of 2 GW, and hence the new nuclear plants were similarly large (or even above 3 GW), and even under the best circumstances their design and construction would have taken most of a decade. But this too had changed. In 1974 the AEC was abolished and replaced by the Nuclear Regulatory Commission, which embarked on a seemingly endless series of regulatory interventions that slowed the new projects while raising their costs. Utility managers, accustomed to count on a virtually guaranteed doubling of electricity demand in a decade and the completion of new large stations in five to six years, now found themselves with steadily declining demand and protracted construction periods, and faced a future in which there might be no demand for electricity generated by plants completed at significantly higher cost after ten to fifteen years under construction.

In many cases, abandoning the ordered plants became the only way out of the dilemma. By 1975 new orders were down to four reactors, but there were thirteen cancellations. The last orders for two new reactors (and fourteen cancellations) came in 1978, and in 1979 the always present public distrust of nuclear electricity generation, the ineradicable legacy of conflating bombs and reactors, admixed with fears of unseen and unfelt radiation, was strengthened by the accident at the Three Mile Island plant in Pennsylvania, though no radiation was released from the plant. The public's unease was only reinforced by the reactor meltdown at the Chornobyl nuclear plant in Ukraine in May 1986. Unlike all the American reactors, the Soviet-designed reactor had no containment building, and radiation spread over a large area of Ukraine, Belarus, and Central and Northern Europe. Thirty-one people were killed almost instantly, 134 people were treated for acute radiation syndrome, and although detailed long-term health appraisals did not show any evidence of a higher overall cancer incidence or a higher mortality among the most affected populations, the accident and the need to entomb the reactor and ensure the structure's integrity for many generations have inevitably undermined the image of fission as a reliable and clean electricity generation alternative.

This effect was mostly felt in Europe. By 1986 the failed prospect for fission was already beyond salvation: during the 1980s there were no new reactor orders in the US, just cancellations, eventually amounting to 120 units. The most famous illustration of the consequences of endless construction delays and huge cost overruns was the collapse of the Washington Public Power Supply System. In 1975 the utility planned to spend $2.5 billion on two nuclear power plants, but in January 1982, after that amount had been spent and it became clear that completing the plants would cost nearly $12 billion, all work was stopped at the two sites and the utility folded on June1983, resulting in the largest municipal bond default in the US history. In 1985 *Forbes*'s review of America's fission experience called the country's nuclear program "the largest managerial disaster in business history, a disaster on a monumental scale. . . . Only the blind, or the biased, can now think that the money has been well spent."

Inevitably, there were major corporate casualties, none more prominent than Westinghouse Electric, the company established in 1886 by George Westinghouse, one of the great pioneers of the early electric era. Between the 1950s and 1990s it became, together with GE, the leading reactor supplier, but cancellations and cost overruns led to its demise. In 1998 Westinghouse Power Generation was sold to Siemens, and the next year British Nuclear Fuels bought Westinghouse Electric Company, Toshiba took over in 2006, but in 2017 the reactor business went bankrupt once more (owing to losses on several US reactors under construction) and the company is now planning a three-way split to be completed by 2024.

What has been the real cost of fission's unfulfilled promise? In 1999 the Nuclear Information and Resource Service concluded that the American nuclear industry got more than 96 percent of all monies, about $145 billion (in 1998 dollars), appropriated by Congress between 1947 and 1998 for energy-related R&D. That would be about $167 billion in 2021 monies, and the opportunity cost of that investment would be now more than $1 trillion. The real cost of the entire fission cycle will not be known for a very long time as it must include the eventual costs of reactor decommissioning and of storing highly radioactive wastes for unprecedented spans of hundreds to thousands of years, creating the problem of de facto eternal vigilance, something that has yet to be solved satisfactorily by any nation!

America's leading role in the development of nuclear fission affected the global trends. In 1973 there were 132 reactors generating about 173 terawatt-hours of electricity. OPEC's move should have benefited the nuclear industry as it offered the rich economies of Europe and North America a reliable energy supply immune to sudden price increases by a production cartel. And indeed, fifteen years later, in 1988, there were more than three times as many reactors, 416, operating worldwide and generating ten times as much electricity (1,727 terawatt-hours) as in 1973. But this was an inevitable consequence of the long time spans needed to complete new stations. Then the big stagnation set in as orders for new reactors nearly disappeared throughout the Western world and additions in Russia and Asia only kept pace with the decommissioning of old units.

In 2020 the world had 443 operating reactors, only about 6 percent more than three decades earlier, but their larger average capacities and higher capacity factors (some reactors now generate more than 90 percent of the time) translated into about 2,500 terawatt-hours of nuclear electricity generation, about a third more than in 1990—but barely more than in the year 2000! And the best forecasts published by the International Atomic Energy Agency show that the planned construction of new reactors (mostly in China) will be just enough to make up for the decommissioning of old reactors (many have now been in operation for more than four decades, much longer than originally intended) and that by 2030 new capacities may just compensate for the retirements that will take place during the 2020s. Even a high-alternative forecast, assuming continuing reactor construction, has the fission capacity at 4.5 percent of the world's total, less than the 5.3 percent in 2019, while the low-alternative forecast sees a more than halving of the current capacity's share, to 2.3 percent.

As for individual nations, the retreat has been the norm. In Europe fission did not meet with universal approval: several countries, including Austria, Denmark, Greece, Ireland, Italy, Norway, Poland, and Portugal, have not built a single nuclear plant. Germany and Sweden decided on an early termination of their generation, and even the most successful national nuclear programs are now in retreat. Following the Calder Hall accident, Britain built twenty-five Magnox reactors by1971 (all now shut down), fourteen advanced gas-cooled reactors (using slightly enriched UO_2 pellets contained in steel tubes), and one pressurized water reactor. Total generating capacity has been declining since 1999. It should reach half its maximum by 2025. Two new reactors are to start working in 2026 and 2027, but their construction, originally proposed in 2011, has been beset by problems.

The French program was by far the most successful Western attempt at converting electricity generation to fission. Électricité de France based it on standardized designs of American pressurized water reactors. It enjoyed broad public approval, and its fifty-nine reactors, sited throughout the country, eventually supplied nearly 80 percent of all electricity in France (a lower share more recently) and made possible considerable sales to neighboring states. But the company placed its last reactor order in 1991. The Soviet nuclear program, forever marked by the Chornobyl meltdown, continued in Russia after the USSR's demise, but by 2020 Russia was producing only about 21 percent of the country's electricity. Japan, devoid of domestic oil and gas, embarked on a major expansion of nuclear energy generation, and eventually derived 30 percent of the total demand from fission. The meltdown of three reactors at the Fukushima Daiichi station in March 2011 led to the complete shutdown of all reactors and halted further nuclear expansion, so that by 2020, fission's contribution was just over 5 percent.

And the breeders? Their development was based on several mistaken beliefs: that uranium-235, a fissionable isotope, was so scarce that its resources could not support large-scale nuclear generation; that technical problems could be overcome within a reasonable time; and that the costs would be competitive. In the US in the early 1970s a big demonstration breeder project was also favored for several political reasons by then president Richard Nixon. It was welcomed by the AEC to justify that organization's continued existence (its original mission had been to provide enriched uranium for nuclear bombs and reactors, which had been accomplished many years before), and it enjoyed considerable congressional support. In 1971, when the funding for the demonstration breeder began, its cost was estimated to be no more than $400 million, with the electric utilities prepared to pay nearly two-thirds of the total. Projected costs had nearly doubled within a year, and by 1981, when the Clinch River breeder became the largest public works project in the US, the total was forecast to surpass $3 billion.

Technical problems with the design and the rising costs of reprocessing (separating plutonium from spent nuclear fuel) sent costs soaring, and by the time the project was canceled in 1983 it had cost $8 billion. That would be about $20 billion in 2021 monies, but this did not deter other countries from continuing to waste additional billions. Most prominently, France completed a large, full-scale breeder (the 1.2 GW *Superphénix* at Creys-Malville) in 1986, but the accident-prone reactor, which was out of commission for extended periods of time, was finally shut down in 1998. Japan's breeder program was terminated on December 8, 1995, after the leakage of nearly 650 kilograms of liquid sodium. Eventually the cost for abandoned breeder developments (also in 2021 dollars) reached nearly $16 billion in Japan, almost $11 billion in the UK, $8 billion in Germany, and almost $7 billion in Italy. Globally, with the addition of Russia, China, and India, the tag would be approaching $100 billion, a sum that illustrates the power of nuclear lobbies and (against all evidence) of the stubborn beliefs of experts advising governments.

By the late 1980s it had also become clear that the second option for the second nuclear era, the deployment of much better designed, smaller, and less expensive but more reliable and inherently safe fission reactors will not happen anytime soon. These designs were mooted even before the US locked itself into submarine-derived pressurized water reactors, and since the early 1980s there have been many more or less detailed conceptual and pilot-plant proposals, particularly for small modular reactors, but no new radical departures. Nor has there been any firm, binding commitment to accelerate the resurrection of well-tested fission capacities as part of a multifaceted strategy to reduce the CO_2 emissions emanating from the world's primary energy supply. True, some formerly vocal opponents of fission have become enthusiastic supporters of large-scale nuclear electricity generation, and some governments—predictably France, but also the US and the UK—have included fission among their choices of techniques for accelerated decarbonization.

In the early 2020s it has become harder to keep up with the news about governments and private investors announcing plans for developing yet more versions of a small modular nuclear reactor, one with less than 300 MW of installed capacity. Interested countries include Canada, China, the Czech Republic, Russia, South Korea, and the US. Rolls Royce, in its first venture into nuclear energy generation, now promises to build eventually up to sixteen such small reactors in the UK. Some of the designs are returning to the use of molten salt, abandoned decades ago: small molten salt reactors are being developed by Kairos Power in California and with government support in China, and in October 2020 the US Department of Energy awarded Seattle's TerraPower funding to demonstrate Natrium, a 345 MW sodium fast reactor with molten salt energy storage.

And then there are the microreactor startups such as Oklo, whose reactor could use spent fuel from conventional large reactors to produce just 1.5 MW of electricity, enough for an industrial site or a university campus. But the company's website shows an image of a large wood-and-glass chalet set against a mountain silhouette, unmistakably signaling the pristine "green" pedigree of a design to be operated without anybody on-site. In light of the past experience with nuclear promises, the only sensible attitude is to wait and see how many of these announced plans will, even with the added incentive of accelerated decarbonization, become actual working prototypes, and then how many of those will make the second cut to lay the foundations of future commercial opportunities. In any case, no nation has announced any specific, detailed, and binding recommitment to what would have to be a multidecadal program of reactor construction.

Perhaps the most unfortunate reality concerning fission's badly missed promises—in 2021 providing about 5 percent of all installed capacity and about 10 percent of all commercial electricity generation, rather than dominating the industry with thousands of reactors, most of them breeders—is that we have no excuse to claim the inevitable ignorance that complicates all new endeavors. The most knowledgeable scientists, engineers, and utility managers were well aware of most of the challenges (from commercializing suboptimal reactor designs to indefensibly optimistic claims of competitive costs), inherent disadvantages, and less than appealing features of the new technique (radiation risks, fear of accidents, the need for lasting vigilance, security concerns, including terrorist attacks on nuclear stations and weapons proliferation). Commercial fission should have been developed more deliberately, more cautiously, and with much more attention given both to its public acceptance and to the eventual long-term storage of its radioactive wastes.

Before leaving this saga of expensive failures I must emphasize that I have assessed their magnitude against the backdrop of vastly exaggerated promises investing fission with the power of transformative salvation. When judged simply by its actual achievements, the post-1945 development of fission has been a "successful failure." I began to use this *contradictio in adjecto* description before the end of the twentieth century, and the past two decades have only confirmed its accuracy. Though the grand promises of a new epoch ushered in by a brilliant high-tech solution failed, nuclear generation has been a partial (if very expensive) success.

On the global level, the judgment is more subjective: is 10 percent of all electricity generated by fission significant or marginal (potentially to be made up by increased conversion efficiencies)? For many nations the benefits have been clear: in 2020 nearly 250 million living in thirteen affluent economies derived more than a quarter of their electricity from fission, and in eleven European nations the share was higher than one-third. Loss of this capacity would be highly disruptive, its gradual replacement costly. No less important, reactors in Europe, the US, and Canada have been operating with impressively high annual load factors—some even in excess of 95 percent, reliably providing the required base load (the minimum generation required without interruption)—and have done so safely, without releasing any greenhouses gases or any harmful doses of radiation while lowering the mortality that would arise from generating similar amounts of electricity by coal-fired stations (because of their emissions of particulates and acidifying gases).

By 2022 there were two new powerful incentives to go nuclear: the quest for accelerated decarbonization of the global electricity generation and Europe's need to reduce its reliance on Russian energy. Still, the response remained unclear. In February 2022, France announced a plan for fourteen new reactors by 2050 (how many will be actually built?)—but in May 2022 Germany refused to prolong the operation of its last nuclear stations past the end of the year. The country was facing the shortage of Russian natural gas, but the ideological zeal overrode the common sense as the ruling coalition's Green Party consented to more coal burning rather than generating more carbon-free electricity! And there are still no clear, binding commitments to building numerous small reactors in the US or Japan. Nuclear realities still keep falling far short of the initial, truly transformative, promise.

SUPERSONIC FLIGHT

While Wilbur watched, Orville Wright made the first powered flight, or rather a short hop of 36 meters lasting twelve seconds, above the sandy beach at Kitty Hawk in North Carolina on December 17, 1903. Then they switched places and completed three more short flights: the last, and the longest one, lasted fifty-nine seconds. Remarkably, almost four years went by before anybody else could fly a heavier-than-air machine for more than a minute. Such were the beginnings of powered flight during the first decade of the twentieth century, and perhaps nothing illustrates better the subsequent speed of aviation advances than the fact that forty years after the first breakthrough, aircraft engineers were beginning to think seriously about designing a plane that would travel considerably faster than sound, shortening trips between Europe and North America to less than the time that elapses between breakfast and an early lunch.

Reciprocating (piston) engines rotating aircraft propellers dominated commercial aviation until the late 1950s, but in 1943 both the UK and Germany were preparing to deploy their first jet fighter planes (the Gloster Meteor and the Messerschmitt 262, respectively, with the Germans being first in combat) powered by turbojet, continuous-combustion gas turbines. While the Mustang, America's most successful propeller fighter, could reach about 630 km/h and the British Supermarine Spitfire less than 600 km/h, the maximum speeds of the two pioneering jet fighters, 970 km/h and 900 km/h, approached the speed of sound. In aeronautics, the Mach number (named after the German physicist Ernst Mach) is the ratio of object speed to the speed of sound. At sea level (and at 20°C), sound travels 340 m/s, or about 1,224 km/h; the speed of sound declines slightly with altitude, and at 11 kilometers above sea level, a typical cruising altitude for jetliners, it is about 295 m/s or 1,063 km/h, and hence a Boeing 787 cruising at 903 km/h will fly at M 0.85. All speeds of $M < 1$ are subsonic; transonic is the term used for speed in the vicinity of M, and the supersonic range is $1 < M < 3$.

With the first jet fighters being almost transonic, it appeared inevitable that the M1 would be surpassed as soon as better engines and better airframe designs become available, and that these advances would be transferred from military to commercial airplanes. That was indeed the case. On October 14, 1947, Chuck Yeager flew the X-1, a rocket plane, at a speed faster than sound, and transonic fighters and bombers then entered air force fleets in the US, UK, and USSR. The first commercial jetliner, the ill-fated British Comet (whose four deadly accidents were not caused by jet engines but by stress around square window frames that eventually led to catastrophic decompression), entered its brief service in 1952 at M 0.7, and the first successful and widely adopted jetliner, Boeing's 707, began its scheduled flights in October 1958 at M 0.83.

Preliminary studies of supersonic flight were done during the early 1950s in the UK, US, and USSR. In 1959 the annual report of the International Civil Aviation Organization acknowledged these developments and noted not only that "there is now general agreement amongst the potential manufacturers on the technical feasibility of producing a supersonic transport aircraft in the relatively near future—that is to say, by about 1965 to 1970" but also that 1959 was "the year in which realization became general that such an aircraft not only was a practical possibility, but almost certainly would be the successor to the present jet transport."

This mistaken belief in supersonic transport as the obvious next step in commercial aviation was promoted (for different reasons) by the governments of the UK, France, the US, and the USSR, and the resulting quest for its realization led to many failures, all costly, some brief, others prolonged. During the late 1950s, Britain and France—having lost their colonial empires, been denied US support for their ill-conceived Suez military action, and relegated to secondary roles by the superpower rivalry between the US and the USSR—were developing, independently, supersonic aircraft designs, and eventually decided to join forces. The formal cooperation treaty was signed on November 29, 1962, and the Concorde venture was launched, intended to reclaim some of the old great power glory. Sud-Aviation and Bristol Aerospace shared the construction of the airframe, and Bristol-Siddeley and SNECMA (Safran Aircraft Engines) developed the engines. Eventually the airframe development phase extended from 1972 to the end of 1978, engine development ended only in 1980, and the production of twenty planes lasted between 1967 and 1979.

Maximum speed was restricted to M 2.2 in order to use conventional aluminum alloys (flights above M 2.2 require titanium and special steels because of thermal limitations). The first test flight of the French prototype was on March 2, 1969; M 1 was reached briefly for the first time on October 1, 1969, and M 2 was sustained on November 4, 1970. Extensive testing of both prototypes followed, and commercial operations began on January 21, 1976, with concurrent flights from London to Bahrain and from Paris to Rio de Janeiro. During the twenty-seven years of their commercial operations, British Airways Concordes flew regularly from London to New York and in winter to Barbados, while shorter intervals of service included Bahrain, Singapore (via Bahrain), Dallas, Miami, and Washington, D.C.'s Dulles airport. Air France destinations included New York and, for shorter periods, Caracas, Mexico (via D.C.), Rio de Janeiro (via Dakar) and Dulles; there were also some three hundred charter destinations all over the world (fig. 3.6).

Eventually, New York remained the only transatlantic destination. On July 25, 2000, a French Concorde taking off from Charles de Gaulle Airport had one of its tires punctured by a piece of metal that had fallen from a just-departed plane. According to the official investigation, this ejected debris ruptured a Concorde fuel tank, and the resulting major fire and loss of engine power led to the deaths of all people abroad (one hundred German tourists and a crew of nine). But, as is common with airline accidents, there were other contributing circumstances, above all the fact that the plane was overloaded and tried to take off with the wind.

Figure 3.6 The Concorde in the British Airways livery. *Source:* Miles Blaine Collection, San Diego Air and Space Museum Archive.

In any case, the catastrophe grounded the airplanes, and the renewed service lasted only until 2003: on October 23, 2003, the last Concorde flight departed JFK for Heathrow.

The Soviet Tupolev Tu-144, an all too obvious derivative of the Concorde (betraying its true origins as a result of extensive Soviet industrial espionage), was an even greater failure. The plane's development was clearly a part of the Soviet effort to demonstrate technical prowess, to add to the regime's record of space firsts (*Sputnik* in 1957, Gagarin, first man in space, in 1961). The plane's design was revealed in 1965 at the Paris Air Show and its prototype was flown on the last day of 1968 to beat the first French Concorde trial, which took place on March 2, 1969. In 1971 the Soviets sent it again to the Paris Air Show where a pilot error caused a spectacular crash. Production ceased in 1982. During the last years of its short service the plane carried largely airmail. Its final flight was in 1984.

Remarkably, Americans avoided their own version of the supersonic failure, but not for any lack of trying. In the early 1960s an American version of supersonic transport (SST) was seen as an inevitability. Because others will build such planes, the US must keep its commercial aviation superiority, recently demonstrated with the sequence of Boeing 707,

727, and 737. This rationale was repeatedly stated by politicians and the plane's promoters: to retain US primacy in aircraft design, not to fall behind such has-beens as the UK and France, not to be bested by the USSR. In a direct response to the Concorde project, President Kennedy announced the development of an American supersonic plane on June 5, 1963, two years after he committed the country to the Moon landing before the end of the 1960s.

The ultimate goal was a lofty one. The Federal Aviation Administration was claiming that the aim of the US program was "a safe, practical, efficient and economical vehicle" and that "we should not go forward, and we do not plan to go forward, unless criteria to meet these objectives are met." Some promise! Industry had no doubt who should pay for it: it was 90 percent government financed, and even congressional leaders were willing to go for a 75–25 percent cost split. Senator Warren Magnuson, a senior member of the aviation subcommittee of the US Senate Commerce Committee who hailed from Boeing's home state of Washington, claimed that the country was "developing an airplane to carry America and the world into the turn of the century."

At that time the consensus among the likely manufacturers was that the first supersonic aircraft, flying closer to 1970 than 1965, could be as fast as M 3, but Kennedy's proposal envisaged a nearly 160-ton plane with the range of 6,400 kilometers cruising at M 2.2 and hence requiring titanium for its construction. Kennedy's message to Congress had also identified the three obvious problems: that the technical challenges posed by the supersonic sped could not be solved, that SST would remain uneconomical, and that the sonic boom would create "undue public disturbance." Eventually, all of these problems manifested themselves, and the weight of this combination led to the cancellation of the government support and hence to the end of American SST.

But it took nearly a decade to get there. In 1967 Boeing's variable-geometry (swing-wing) design was selected over Lockheed's conventional configuration, but after a year of trying, Boeing abandoned the design. The Federal Aviation Administration then chose a larger, 340-ton version, as heavy as the Boeing 747 and twice as large as originally intended (fig. 3.7). But the late 1960s were a time when environmental impacts, from pollution to noise, were attracting public concern, and SST became

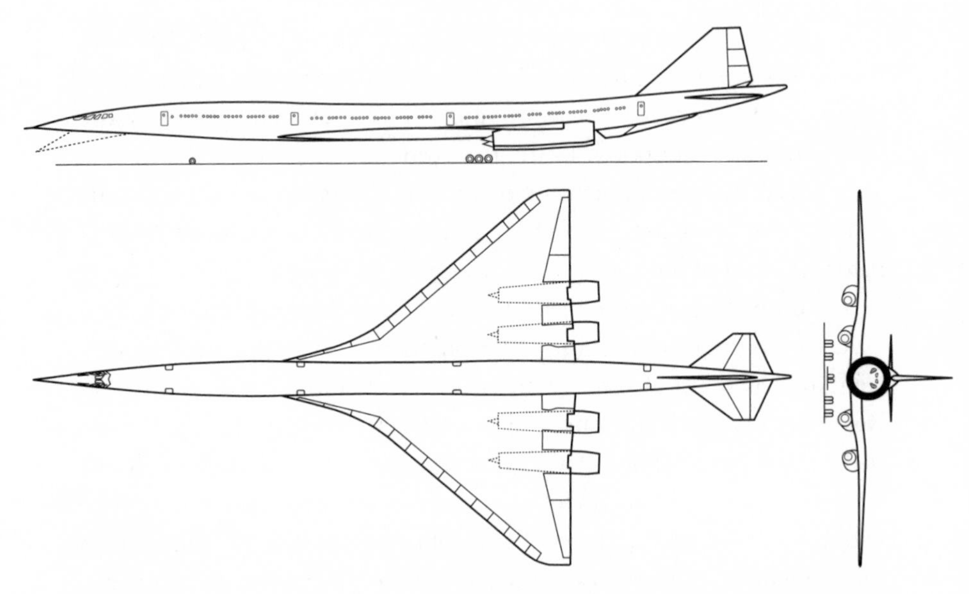

Figure 3.7 Three views of the Boeing 2707–300, the SST craft that never took off. *Source:* Illustration by Nubifer, https://commons.wikimedia.org/wiki/File:Boeing_2707-300_3-view.svg, licensed under CC-BY-SA. Reprinted with permission.

environmentalists' early and prominent target. Between 1967 and 1971 the public campaign against sonic booms became vocal, well publicized, and politically more influential. In 1969 two reviews of the project ordered by the newly elected president, Richard Nixon, concluded that because of the excessive costs of the project and the "intolerable" effects of sonic booms the government should withdraw its support.

But in September 1969 Nixon decided to go ahead with the project, and the contest shifted to Congress. Expert witnesses at congressional hearings detailed all the downsides, from poor efficiency and limited range to unjustifiable expenditures and extraordinarily high levels of noise. Physicist Richard Garwin added another accomplishment to his list of achievements (from working on the detailed design of the hydrogen bomb to computer printer development): while still serving on the President's Science Advisory Committee (PSAC) he became perhaps the SST's most authoritative and most effective critic. Finally, on March 24, 1971, the Senate voted 51–46 to end further funding of SST, and Nixon disbanded the PSAC after his reelection (his unhappiness with Garwin's opposition was widely credited for his doing so).

Why have all of these ventures ended in failure? America lacked what Europe had to pursue this expensive, unnecessary, uneconomical, and unjustifiable project: cooperation (if not collusion) among decidedly more dirigiste governments, national airlines, and government-supported aircraft manufacturers prevailing in the absence of public dissent. But that served the US well as it spent "only" about a billion dollars on the failed attempt to maintain America's illusion of aviation primacy. Instead, America's strategic thinkers would have earned their keep had they been more concerned about the establishment of Airbus Industrie on December 18, 1970, with France, Germany, and the UK getting together to produce new commercial jetliners. That move eventually led to America's second-best status: during the second decade of the twenty-first century Airbus received more orders for new jetliners than Boeing in all but two years.

But the two supersonic successes, getting the Concorde and the Tupolev off the ground and into commercial service, were in reality no such thing, just more or less protracted and much more expensive failures. Why did the faster flight, even when promoted and financed in unprecedented ways, not turn out to be a natural successor of the now more than sixty-year-old subsonic aviation? Why we have not seen a second wave of supersonic airplanes? These questions have always had a number of clear, convincing answers, and the failures could have been (and indeed were) predicted by critical commentators even when the enthusiasm for national projects was at its highest during the 1960s. Moreover, most of the reasons for past failures have not been eliminated or resolved, and they will have to be faced by the most recent attempts to reintroduce supersonic flight.

Four fundamental constraints are apparent: a plane design dictated by the need to overcome enormous supersonic drag, engines powerful enough to sustain M 2, accomplishing this economically, and doing so with acceptable environmental impacts. The lessons of the Concorde are an obvious start of these explanations. The planes looked streamlined and graceful on the tarmac and in flight. They could fly just a shade faster than M 2, and hence they could make it from London to Washington, D.C., in less than four hours: the arrival time in America was thus ahead of the London departure time. These were much admired facts, but just about everything else about the plane was notable because of the Concorde's problems and negatives stemming from the imperatives, and hence unavoidable constraints, of supersonic flight. The most fundamental of these taxing requirements is to overcome the increased drag with greater propulsive force.

Drag coefficient (a dimensionless ratio of drag force and the product of air density, the square of speed, and the object's surface area) peaks at just above M 1 and is lower at both subsonic and supersonic speeds. That is why all modern jetliners cruise at about M 0.85 and why their speed has remained fairly constant since the Boeing 707 began to fly in 1958. But the lift-to-drag ratio (L/D)—and hence the range of an aircraft—decreases with speed: for the Boeing 787, cruising at M 0.85, it is 18, at M 1 it is about 15, at M 2 just 10. And while the Boeing 787 has a maximum range of nearly 14,000 kilometers, the Concorde could go only less than 6,700 kilometers, not enough for a transpacific flight without refueling (the flying distance from San Francisco to Tokyo is 8,246 kilometers).

To minimize the drag coefficient, it is necessary to keep the plane's area (that is, the diameter of its fuselage) as small as practically possible: a slender fuselage is a must, going against the trend of wider bodies on the favorite subsonic planes. The Concorde's fuselage had a diameter of just 2.9 meters, about 20 percent less than the Constellation, the largest long-range piston-engine plane of the pre-jet era and only about half that of the Boeing 747 or the latest 787 (5.77 meters). As Richard K. Smith has remarked, "The 747 transformed the Concorde to a claustrophobe's horror." Concorde's single-aisle two-by-two seating had adequate legroom but restricted elbow room; the seats were padded, but the cabin had the ambience of a crowded economy charter flight.

But even with its small cross section, to withstand higher speeds the Concorde's mass was higher than that of a comparably sized subsonic plane, and the plane had a relatively low payload, only about 10 percent of its gross weight, half the rate for the Boeing 747. Supersonic planes cannot earn money by carrying cargo, while every wide-body jetliner is also a significant commercial cargo carrier, a reality you can see from your window seat above the cargo door or from a terminal, watching larger palette trucks loading the bellies of passenger planes. The material requirements to build airplanes are more exacting as the speed increases, but up to M 2 they can be largely met with the best aluminum alloys. At M 2.2 leading edges have temperatures as high as 135°C, higher than the temperature limits of the fiber-reinforced polymers (90°C) that now make up most of the fuselage and wings in the latest jetliners. Heavier titanium and steel are the most obvious options (polymers have higher tensile strength per unit of mass, but some steel alloys are good up to 800°C).

And supersonic planes cannot take advantage of the most efficient modern high-bypass-ratio engines where only a tenth or even less of the turbofan-compressed air moves through the turbine, the rest bypassing the core and thereby boosting fuel efficiency while also lowering engine noise. Moreover, the Concorde's engines needed afterburners to provide the thrust needed for takeoff and for pushing the plane through the peak drag transonic zone, but afterburners boosted fuel consumption, complicated the already expensive maintenance, and increased the takeoff noise. The Concorde burned more than three times as much kerosene per passenger as the first wide-body Boeing 747. That made less difference in 1970, when crude oil sold for $2 a barrel, but a decade later, after two episodes of oil price hikes by OPEC, oil was selling at close to $40 a barrel.

Profitable supersonic flight looked unattainable even according to the earliest, overoptimistic estimates. To begin with, during the late 1950s and the early 1960s major international airlines faced financial problems as a result of their rapid switching to jets before their most recent long-range propeller planes (Lockheed's Constellations, the DC-7, Britannia) were paid for. Just a decade later they faced an even greater dilemma: to acquire a fleet of brand-new wide-body jetliners (the Boeing 747 entered service in 1970, the McDonnell Douglas DC-10 in 1971) or to wait for the first supersonic airplanes, with the latter choice made even more uncertain by the likelihood of the first generation of aluminum supersonic planes (maximum M 2) getting displaced by still to be developed nonaluminum supersonic planes (speeds up to M 3). In 1965 an estimate for the fixed cost of transcontinental US flights (never mind that the sonic boom prevented them) was about four times the analogous rate of subsonic planes, with the variable costs about equal and the maintenance labor costs four times higher.

Because of the enormous costs of its development—the best estimates are that the eventual unit cost was twelve times the original number—and the limited number of planes in service, the Concorde could have never made any profit, but OPEC made its losses much worse. In contrast, to say that the Boeing 747, whose first flight also took place in 1969, revolutionized global passenger aviation is just to state an indisputable fact. Airlines found it highly profitable, passengers liked the low prices and the roominess the wide-body design could offer, and by 2022 Boeing built nearly 1,600 747s. In contrast, only twenty Concordes were ever built, only fourteen entered commercial service, and only Air France and British Airways "purchased" them, with the acquisitions and every flight heavily subsidized by French and British taxpayers.

And even if, miraculously, supersonic flying had somehow come closer to profitability, the environmental limits on routing and destinations would have set it back again. Richard Garwin illustrated the effect of the plane's sonic boom by equating its peak intensity to "the simultaneous takeoff of 50 jumbo jets"—and there is no public constituency for that kind of repeated experience. As a result, it was clear that even if it eventually entered commercial service, the American SST would never fly across the continent, and the Concorde's New York landings were resisted, denied, litigated, and eventually permitted (with conditions) only after years of court battles.

Supersonic flight did not become the next step in the "natural" sequence of steadily rising transportation speeds: since the late 1950s these speeds have remained constant at M 0.85. The best appraisal of the quest for supersonic speeds was published by Richard K. Smith, an American aviation historian, who called it the "frenzied international aeronautical saga of communicable obsessions": "From the start to finish, in Britain, France, and the United States, the supersonic airliner was a flying machine that the world did not need; it was a political airplane."

But the conviction that faster speed is in the natural order of things is still around, and I must close the story of supersonic flight by looking at recent attempts at its resurrection. Half a century after the US Congress killed the American SST aircraft and some two decades after the Concorde's last flight, there are new supersonic dreams. Their exaggerated claims, super-optimistic timetables, and true-believer convictions regarding the imminent solutions to all technical problems are highly reminiscent of the early 1960s, but this time it is not the collusion of European governments and airline and aircraft companies but an American startup that comes with the most stunning promises.

The EU, with its green preoccupations and its preference for stringent regulation, does not appear to be interested in reliving another Concorde experience. In Russia, the Central Aerohydrodynamic Institute says it is designing a supersonic airplane (M 1.6, sixty to eighty passengers, takeoff-weight 120 tons, range 8,500 kilometers) made of composite materials and with a sonic boom reduced to 65 dB, puts its production start in 2030, and it expects domestic demand of twenty to thirty planes a year. And Tupolev's design bureau hopes for a second run, working on a business jet (M 1.3–1.6, thirty passengers), with the first flight promised in 2027.

Before you take any of this seriously, consider what success Russia had with its Sukhoi Superjet, an ordinary narrow-body regional jetliner intended to compete with the ubiquitous Airbus planes. Sukhoi Aviation, the country's famous designer of supersonic jetfighters (the Su-30 flies at M 2), began its development in 2000 and the first commercial flights came in 2011, but by 2020 the Mexican Interjet was the only non-Russian airline to have placed a small order (and endures maintenance problems with idled planes). The same skepticism should apply to recent American plans, but before COVID-19 came around there appeared to be some strength in numbers: in 2019 there were four American companies developing supersonic planes: Aerion, Spike Aerospace, Lockheed Martin, and Boom Technology.

Aerion's Supersonic, established in 2004, was to have its business jet (eight to twelve people, M 0.95 over land, M 1.4 over ocean) flying by 2023 and in service by 2025. The company had partnership agreements with Boeing and General Electric and expected sales of five hundred to six hundred aircraft during the next twenty years. In May 2021, after seventeen years and not even a prototype model produced, the company folded. Spike Aerospace says on its website that it is developing an "ultra quiet supersonic business jet" for eighteen people that will fly at M 1.6 "without creating a loud sonic boom." The record so far: the first supersonic flight of the forty- to fifty-passenger design was to be in 2018, with certification by 2023, then 2025, followed by a switch to an eighteen-passenger design to fly in early 2021, with deliveries planned for 2023. The reality at the end of 2021: nothing in the air.

That leaves Lockheed and Boom Technology. Lockheed's plans for its M 1.8, forty-passenger twin-engine jets are vague, with progress depending on the success of the X-59, NASA's experimental supersonic prototype that the company has been building since 2018. In any case, Lockheed believes the supersonic plane will need a new engine, and it has no timeline for the plane's introduction.

In contrast, few chief executives have been so boastful or issued so many timelines as Blake Scholl, the founder and CEO of Boom Supersonic, a private company that plans to build the Overture, an M 2.2 plane for fifty-five people. In 2019 Scholl had the commercial service beginning in the mid-2020s, with a head-spinning market order estimate of one thousand to two thousand planes during the first decade of production.

In October 2020 the company rolled out the XB-1, a one-third scale model of the Overture that is to take off in 2022 to prove the basic design, cockpit ergonomics, and "even the experience of flight itself." But that experience will be limited to a single pilot, and the plane will be powered by three small General Electric J85 turbojets that hardly need to prove anything after more than half a century of military and civilian service (it was designed in 1954). Obviously, the full-size plane will have to be powered differently, and Rolls-Royce was enlisted to do that, but no engine has been selected. In 2022 Boom's timeline was as follows: the company announced it would build a new factory in 2022, with construction of the first Overture plane starting in 2023; the first plane roll-out in 2025, the first flight would in 2026, and, after a quick certification, the plane with sixty-five seats would enter commercial service in 2029.

This means that a company that has never built a single passenger jetliner intends to design, secure complex supply chains for (modern jetliners are made from parts delivered by numerous specialized subcontractors), assemble, test, and get certified a brand-new supersonic aircraft in less time than it took Boeing, the world's leading aviation company and one that has built tens of thousands of planes, to get its latest design iteration, the 787, into service. As Boeing's certification statement says, "The eight-year certification process for the 787 was the most rigorous in Boeing's history, and the design of the 787 incorporates nearly a century of aviation learning and safety improvements." And, as is well known, Boeing still had problems when the 787 began flying.

But Boom, without any experience and with an unprecedented design, intends to do it faster than the world's most experienced aircraft builder. And to top it all off, its planes will be sustainably fueled by zero-carbon liquids. According to Scholl, "What you're basically doing is sucking carbon out of the atmosphere, liquefying it into the jet fuel, then you put that in the airplane. . . . You're just moving carbon around in a circle." Why are not all airlines doing that already? Might it be because such a process is not available to produce aviation fuel on a large scale, and because its best, small-scale versions make a fuel that is at least five times as costly as kerosene? And because using aviation biofuel (which would not be carbon-free unless all field machinery were to be energized by renewably generated electricity) would not be that much cheaper, at least three to four times the cost of kerosene? And because using such fuels in an aircraft that would need at least four to five times as much energy per passenger than the Boeing 787 has no prospect of economic justification?

No matter. In a 2021 interview, Scholl claimed that the ultimate goal was to fly "anywhere in the world in four hours for 100 bucks." He qualified that by saying that that would apply to "two or three generations of aircraft down the line," but even so, it would need a concatenation of many incredibly extraordinary things to happen. "Anywhere in the world" would mean a maximum distance of 20,000 kilometers; in four hours that is 5,000 km/h, or (when cruising at 20 km in the lower stratosphere) M 4.7. That is a lot faster than the fastest military jet ever built, the Lockheed Blackbird SR-71, which could do M 3.2 at 25 km (the much faster X-15 could not take off on its own; it was essentially a rocket dropped from a large plane). All too obviously, all these claims sound too good to be true.

Whatever you may hear (or not hear) about Boom's progress, the basic facts remain. Supersonic flight did not displace subsonic aviation; it has not taken even a small market share from it because, for many reasons, it is not an inevitable next step in airplane development and because its few advantages cannot outweigh its many drawbacks. This reality is not going to change anytime soon.

4

INVENTIONS THAT WE KEEP WAITING FOR

All too obviously, listing even the most notable desiderata in this enormous category of unrealized inventions that we keep waiting for could take pages, but the purpose of this topical chapter is different. I am not interested in recounting such universally desired yet prominently missing advances as eliminating cancer or radically extending the human life span because such goals are based on false premises. In the US the National Cancer Act of 1971—passed during Richard Nixon's first administration and widely dubbed the beginning of the "war on cancer"—was only the beginning of setting up dubious goals. More than half a century later cancer remains the second largest cause of death in the United States—but, at the same time, our record cannot be seen as a serial failure, as a helpless wait for real breakthroughs to come.

When the challenge is viewed, as it properly should be, in terms of specific malignancies, we have achieved many successes, ranging from impressively high cure or remission rates for childhood leukemias to no less impressively improved early detection rates and effective treatment of prostate cancer. And the trend has been moving in the right direction even in terms of overall population-wide incidence and survival. The 2021 annual report prepared by the National Cancer Institute shows a declining mortality from eleven of the nineteen most common cancers in men (including melanoma, lung, leukemia, and myeloma) and fourteen of the twenty most common cancers in women, also led by declines in melanoma and lung cancer rates.

Similarly, the failure to achieving any radical extension of the human life span should not be seen just as yet another unfulfilled promise, as a breakthrough we keep waiting for. Most of the world's affluent countries have actually achieved an impressive extension of the average pre-industrial life span, doubling it from about forty years in 1800 to about eighty years by the year 2000, with the highest combined (male and female) longevity now at eighty-five years in Japan. And the way we have done this gives us a fundamental insight into possibilities and limits. If humans had no upper limit to their survival, then the highest survival gains should be observed among ever-older age categories. This was true until the early 1990s, but afterward the gains began to plateau, then declined after reaching one hundred years, and we have yet to see anybody outliving the record-holder, who died at 122.4 years in 1997. Consequently there is a strong case for concluding that the maximum human life span is limited by natural constraints, with the aging of elastin, the protein essential for the operation of internal organs and muscles, being a prominent component of this process.

In contrast to such misunderstood or unattainable goals, I focus in this chapter on the ever-receding fulfillment of much more narrowly circumscribed (if very challenging and exceedingly complex) technical pursuits. Again, I will take them up in their chronological order, starting with the more than two centuries old quest for rapid transportation of people and goods inside evacuated tubes. During the nineteenth century this remained just a fascinating, and only theoretically attainable, idea, but in the twentieth century new materials and new means of propulsion turned it into a still daunting but eventually achievable challenge.

Unlike rapid transport in a (near) vacuum, the second yet-to-be-achieved breakthrough I describe in this chapter—our attempts to make cereal plants (wheat, rice, corn) act as legumes and secure most of their nitrogen requirements by symbiosis with bacteria rather than needing heavy doses of synthetic fertilizers—has not attracted recurring and uncritical media attention. This is not at all surprising because it concerns a more efficient and environmentally less damaging cultivation of staple cereal crops, a topic that is far from the purview of the modern infatuation with everything fast-moving or electronic. Moreover, only a small minority of people in affluent countries understands the nutritional requirement of crops as most people have no idea how much nitrogen, the essential plant nutrient whose availability is the most common limiting factor in crop production, is needed for a good harvest of wheat (to be milled to produce flour for bread and pastries) or corn (to be fed to animals, which produce milk, meat, and eggs).

This possibility arose more than 130 years ago when symbiotic *Rhizobium* bacteria, living in small nodules attached to roots of leguminous plants, were identified as the providers of nitrogen. Prospects for mimicking this arrangement in cereals seemed to rise exponentially starting in the 1970s with advances in genetic engineering, but half a century later—and despite considerable progress in sequencing the bacterial genomes and identifying the genes responsible for imparting the capability of turning inert atmospheric nitrogen into reactive and water-soluble ammonia—we do not seem to be close to having self-fertilizing staple grain crops.

The last breakthrough with repeatedly delayed deadlines for convincingly demonstrating the commercial capability of a possibly epoch-making invention concerns controlled nuclear fusion. This conversion, replicating some of the reactions that keep stars emitting enormous quanta of energy across the span of billions of years, became conceivable even before the first successful demonstrations of nuclear fission. Its first, destructive application was achieved after a remarkably brief period of intensive R&D during World War II, and its first commercial deployment for electricity generation followed a decade later.

But even as the first fission plants were entering regular service during the 1950s, researchers not only began to explore the possibilities of controlled fusion but introduced a design of an experimental device that, more than six decades later, still offers the best hope for achieving the conditions required for sustained, controlled nuclear fusion. But how soon this design, which has been tested in larger and more capable configurations and is soon to undergo what was planned to be the last pre-commercialization proofs of the entire concept, will become available for commercial deployment still remains a matter of educated guesses rather than of confident predictions.

TRAVEL IN A (NEAR) VACUUM (HYPERLOOP)

On August 12, 2013, Elon Musk, at that time the chairman of Tesla, released his Hyperloop Alpha paper. At its very beginning, when outlining

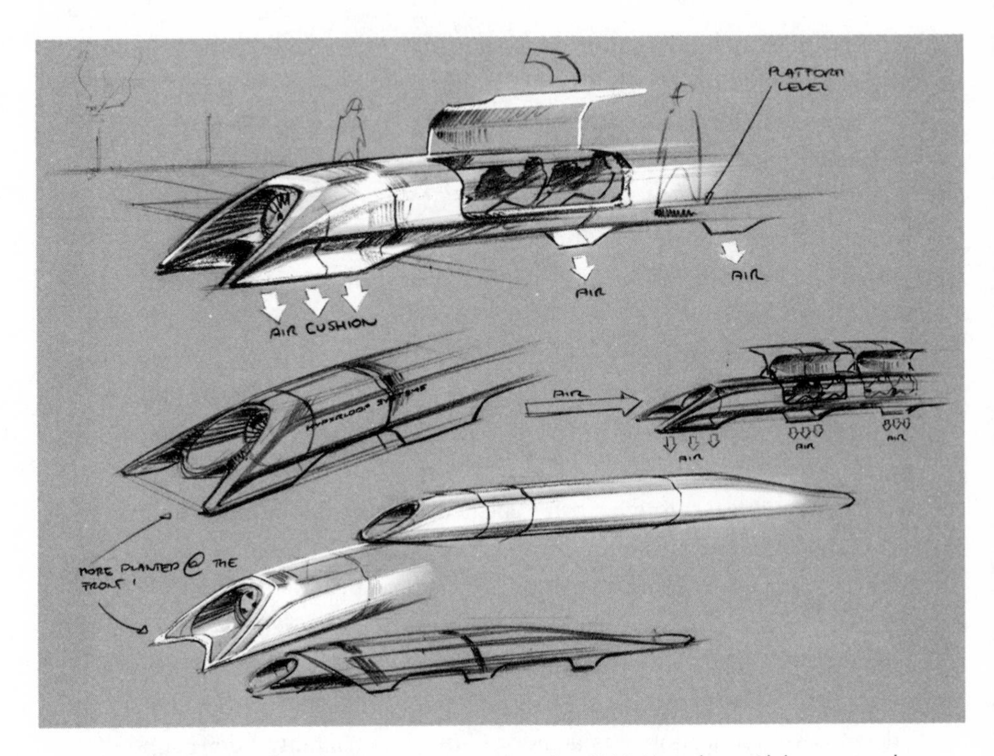

Figure 4.1 The first drawing accompanying the 2013 Hyperloop Alpha proposal. *Source:* Drawing is available at Hyperloop Alpha, http://tesla.com.

the background of the idea, he asked whether there was “a truly new mode of transport—a fifth mode after planes, trains, cars and boats”—that would be safer, faster, cost less, and be more convenient while being immune to weather, sustainably self-powering, resistant to earthquakes, and not disruptive to people living along its route (fig. 4.1). He noted that “many ideas for a system with most of those properties have been proposed and should be acknowledged, reaching as far back as Robert Goddard’s to proposals in recent decades by the Rand Corporation and ET3. Unfortunately, none of these have panned out.” The second sentence of this quotation was quite correct; the first one greatly understated the origins of the idea, and hence the time that has elapsed since its first coherent formulation, a reality that should also lead to exceedingly cautious and highly skeptical appraisals concerning its near-term commercial realization.

But first I must note Musk's wrong label: a loop is a shape that is produced by a curve that bends around and crosses itself, and I leave it to your imagination what shape it would take to make it a hyper (super, excessive) loop. Hyperloop is thus an incorrect—indeed, a highly misleading—term for the means for rapidly transporting people enclosed in capsules (pods) supported on a cushion of air (other designs use magnetic levitation) inside a very low-pressure (near vacuum) and overwhelmingly straight metallic tube (aboveground or in a tunnel) and moved by a magnetic linear accelerator fastened along the way and energized by solar panels placed on top of the straight tube's trajectory (other designs use different prime movers). The misleading classifier aside, this fifth mode of transportation consists of several distinct components whose specific attributes can vary.

The visible infrastructure is a tube with a diameter only large enough to accommodate the pods capable of carrying a small number of passengers. This tube can be built (most economically from prefabricated sections) on pylons aboveground or it can be placed in an underground tunnel. The pod size depends on the number of people (Hyperloop Alpha specified twenty-eight; other designs range between four and one hundred) and on their accommodation: comfortably seated in reclined chairs or supine. High speeds—ranging from subsonic to near sonic (the speed of sound is 1,235 km/h)—are achievable only in a complete vacuum (too costly to attain and maintain on the scales required) or in very low-pressure atmospheres (easier to sustain, but still presenting operational challenges). Hyperloop Alpha specified an internal pressure of 100 Pa, that is, less than 1/1,000th of the pressure at sea level. Pods can ride on air cushions or be magnetically levitated. Modern systems would be energized by advanced linear motors.

The historical record shows that there is nothing new about any of these ideas, that the basic concept for the fifth mode of transportation has been around for more than two hundred years, and that during the intervening time various patents were filed, several detailed proposals were made, and some models and mock-ups of specific components were built. And yet not a single (near) vacuum- or low-pressure-tube, superfast transportation project (be it for people or goods, or both) has been completed and put into operation, not even a trial short-distance link encompassing all of the design's basic components.

A tube is the component with the longest history, but the idea of using very low pressure is also more than two centuries old. Remarkably, the proposals for both these key features of the supposedly revolutionary fifth mode of transportation are older than the Liverpool and Manchester Railway, the first steam-powered intercity transportation conveyance, which began to carry passengers and freight in 1830. George Medhurst, an English clockmaker and inventor, was the pioneer and determined proponent of rapid travel in tubes. In 1810 he published a brief pamphlet titled *A New Method of Conveying Letters and Goods with Great Certainty and Rapidity by Air,* proposing to send letters in small hollow vessels propelled by air pressure in tubes (generated by steam engines), and concluded that the same principle (with commensurately raised pressure) could be used to move goods at speeds at least ten times those achievable with canal or wagon travel.

In 1812 he presented the more detailed *Calculations and Remarks, Tending to Prove the Practicality, Effects and Advantages of a Plan for the Rapid Conveyance of Goods and Passengers Upon an Iron Road Through a Tube of 30 Feet in Area, by the Power and Velocity of Air*, and he revisited the proposal once more, in 1827 (the year of his death), in a publication with an even longer title: *A New System of Inland Conveyance, for Goods and Passengers, Capable of Being Applied and Extended Throughout the Country; and of Conveying All Kinds of Goods, Cattle, and Passengers, with the Velocity of Sixty Miles in an Hour, at an Expense That Will Not Exceed the One-Fourth Part of the Present Mode of Travelling, Without the Aid of Horses or Any Animal Power.*

These short pamphlets were not widely known, but in 1825 the British public could read about a much bolder proposal for using tubes, vacuum, and high speeds to cover the distance of just over 600 kilometers between London and Edinburgh in five minutes (yes, minutes, not hours). The proprietors of the newly formed London and Edinburgh Vacuum Tunnel Company, after "having carefully matured their plans," published (in the *Edinburgh Star*) their prospectus for a joint stock project "with a capital of Twenty Millions Sterling, divided into 200,000 shares, of £100 each, for the purpose of forming a Tunnel or Tube of metal between Edinburgh and London, to convey Goods and Passengers between these cities and the other towns through which it passes."

Boilers would be placed every two miles along the two side-by-side tunnels (tubes), and the steam they generated would be used to create a vacuum. When the vacuum seal was broken right behind the train at the departure end, the inrushing air would instantly propel the train into the tube by pushing on "a very strong air-tight sliding door, running on several small cylindrical rollers, to lessen the friction." The train would carry only goods because the tube would be just four feet (1.2 meters) in diameter, and passengers would be seated in railway carriages running on rails fastened to the tube's top and coupled by strong magnets to the freight train inside the tube whose rapid progress would drag on the passenger train, covering nearly 800 kilometers in five minutes.

The *London Mechanics' Register*, a new periodical established to diffuse scientific knowledge "among the operative classes of society," reprinted the notice in order "to throw ridicule upon some of the preposterous plans now before the public for the investment of money." Precisely! The country's unfolding steam-based industrialization offered many new opportunities for outrageous claims, financial scams, and false prophecies of technical miracles, and the decade's leading satirical illustrator did not miss his opportunity to lampoon the early promise of travel in a vacuum. William Heath (1794–1840) initially called himself a "portrait & military painter," but during the 1820s he published many satirical colored etchings, often alluding to political affairs of the day or lampooning generic human follies.

In 1829 Thomas McLean in London published Heath's colored etching *March of Intellect. Lord how this world improves as we grow older.* The busy image is crowded with such would-be futuristic contraptions as a suspension bridge from Cape Town to Bengal, a four-wheeled steam-powered horse called VELOCITY, a gun-carrying platform that was lifted by four balloons, and a large winged flying fish crammed with convicts being transported from England to Australia. But the etching's center of interest is a large seamless metallic tube that is conveying passengers from Greenwich Hill (in East London) directly to Bengal, thanks to the innovative acumen of the Grand Vacuum Tube Company (fig. 4.2).

By the time Heath pictured in color the intercontinental Britain-India conveyor, enough was known about vacuum to realize that it would be the best option for attaining unprecedented speeds inside a tube, but the

Figure 4.2 Grand Vacuum Tube Company Direct to Bengal: William Heath's 1829 colored etching was in reaction to a less ambitious, but still impossible, project of using that technique to move people between London and Edinburgh. *Source:* William Heath, *A Futuristic Vision* (etching) (London: Thos. McLean, ca. May 1829), Wellcome Library no. 37252i, available at https://wellcomecollection.org/works/re2aprgu. Reprinted under Creative Commons Attribution International 4.0 license.

material requirements made any realization highly premature. In the 1820s there was plenty of cast iron but no mass production of affordable high-tensile steel (material available in bulk only after the invention of Bessemer's converter, patented in 1856) to build such a tube, no reliable means to create and sustain very low pressures inside tubes extending for hundreds of kilometers, and no ready means to enclose people safely in vacuum-enveloped containers.

The decades following the rapid demise of the notion of a five-minute trip from London to Scotland saw assorted proposals, exploratory rail schemes, and even some actual projects involving unusual modes of propulsion, above all attempts to commercialize "atmospheric" railways.

These railways did not need any locomotives and relied on air pressure to push freight cars along the rails. An airtight pipe with a piston was laid between the rails; steam engines situated along the track pumped the air out of the pipe in front of the piston, creating a partial vacuum; and the higher air pressure behind the piston propelled railway cars (connected to the piston by a metal plate protruding through a slot at the top of the pipe). Obvious advantages were the absence of noise, smoke, and sparks from locomotives, as well as the ability to climb steeper grades than locomotive-driven trains.

These efforts began with a proposal for the National Pneumatic Railway Association in 1835. In 1839 Jacob and Joseph Samuda conducted trial runs on a short track, reaching maximum speeds of 48 km/h and a 50 percent vacuum, and in the early 1840s the first commercial line, the Kingstown and Dalkey Railway, operated briefly in Ireland. These trials impressed Isambard K. Brunel, perhaps the country's most famous engineer, so much that he pushed (against the warnings of his engineering peers: Robert Stephenson, the country's leading locomotive designer, called it a "great humbug") its installation on a fifty-two-mile section of the South Devon Railway between Exeter and Plymouth. Work began in 1844, and even before it was completed Brunel had installed an atmospheric railway on a shorter part of the Croydon Railway.

But by September 1848, after less than a year of "atmospheric" operation (steam locomotives were used until 1847 as the system kept malfunctioning) and after a substantial monetary loss, it was all over. For months, Brunel kept promising success, but the lines were plagued by too many insurmountable problems. Perhaps the trickiest part was the moving slot in the pipe: it required an airtight seal to maintain a partial vacuum in front of the piston, but the tallow-treated leather flap, even when not chewed on by rats, provided a poor seal and kept drying out and turning brittle. Additional short-lived (and short-distance) atmospheric railways ran between 1847 and 1860 near Paris, at London's Crystal Palace in 1864 (just 550 meters), and under New York's Broadway between 1870 and 1873 (a pneumatic subway track of a mere 95 meters). More powerful and more efficient) steam locomotives, and before the century's end also new electric traction, made all unwieldy "atmospheric" projects clearly uncompetitive.

The next important development in the long-running saga of tube-enclosed rapid transport came with the proposal for magnetic levitation. The first patents for specific components of this new technique were awarded in 1902 to Albert C. Albertson and to Alfred Zehden in 1905, and at least three inventors contributed to advancing the concept of maglev transportation. Chronologically, the first description came from Robert Goddard, the physicist who became later well known as the founder of American rocket propulsion. During his freshman year at the Worcester Polytechnic Institute, his class was given an assignment on the topic of traveling in 1950, and Goddard outlined his idea of a levitated train inside a tube with propulsion provided by direct current magnets traveling from New York to Boston in ten minutes. He read the project's description to his fellow students on December 20, 1904, and in January 1906 he rewrote it in a form of a short story titled "The High-Speed Bet" and submitted it for publication in *Scientific American*.

The story was eventually condensed, to concentrate on the basic technical facts, and got a mere third of a page in the periodical's November 20, 1909, issue. But even with this delay, Goddard, as he stressed later, had his idea published before Émile Bachelet, a French electrician who emigrated to the US in the early 1880s and filed for a patent on levitated high-speed trains on April 2, 1910. But it was Bachelet's work, not Goddard's proposal, that received unusually extensive public attention. Bachelet was granted the US patent for a "Levitation transmitting apparatus" on March 19, 1912, and his subsequent presentations of a working small-scale model of magnetically levitated train with a tubular prow, powerful "repelling magnets" at the track's bottom, and tubular steel cars on an aluminum base were well received both by invited experts and by print media (fig. 4.3).

The third inventor with pre–World War I magnetic levitation designs was Boris Petrovich Weinberg, head of the Physics Department at the Tomsk Institute of Technology in Siberia. Between 1911 and 1913 he built a model consisting of a 10-kilogram iron carriage, a 20-meter-long (32-centimeter-diameter) evacuated ring tunnel of copper and a series of sequentially activated solenoids on top of the pipe suspending the carriage that eventually circulated at 6 km/h. This proof of concept was followed by proposals for a full-scale project operating at speeds of 800–1,000 km/h

Figure 4.3 Émile Bachelet and his working model of a magnetically levitated railway. *Source:* Émile Bachelet Collection, Archives Center, National Museum of American History.

with passengers lying in cigar-shaped (0.9 meters in diameter, 2.5 meters long) steel cylinders supplied with oxygen. His book, *Motion without Friction,* was published in Russia in 1914, and after he was sent to the US by the Russian military to secure deliveries of artillery shells, brief illustrated descriptions of his proposal also appeared in two American journals, in 1917 in the *Electrical Experimenter* ("Traveling at 500 Miles Per Hour in the Future Electric Railway") and in 1919 in *Popular Science Monthly* (subtitled "An Electromagnetic Method of Transporting You through a Vacuum from New York to San Francisco in Half a Day").

In 1920 Robert Ballard Davy was granted the US patent for a vacuum railway "comprising generally, a tube with stations at intervals, the tube between the stations having a partial vacuum produced therein so that suitably propelled cars moving therein may travel with greater speed by reason of the lessening of the air resistance." Nothing new here, so he also claimed "a novel arrangement in the stations, whereby the car has egress and ingress to the adjacent vacuum tube portions, without admitting enough air to said tube portions to destroy the vacuum," as well as "a novel locking arrangement for the sliding and hinged doors which form important parts of the aforesaid stations."

None of these endeavors resulted in any practical results, but Goddard's proposal received much delayed attention after World War II. Less than three months before his death on August 19, 1945, Goddard applied for the US patent for a vacuum tube transportation system, and on June 20, 1950, the patent, accompanied by three pages of detailed technical illustrations, was granted to his wife, Esther, jointly with the Guggenheim Foundation. But the 1950s were the era of oversized cars, expanding flying, and declining train ridership (in the US it had already peaked in 1920), and though there were additional levitation-related patents issued during the 1950s and 1960s, another notable American proposal came only in 1972 when Robert Salter, at the Rand Corporation in Santa Monica, came up with a very high-speed transit system concept whose "tubecraft" would ride on, and be driven by, electromagnetic waves generated by pulsed or oscillating currents in electrical conductors forming the "roadbed" structure of an evacuated "tubeway."

Incredibly, Salter maintained that the speeds required for his proposed continent-spanning link (New York to Los Angeles) would "certainly be on the order of thousands of miles per hour," and such supersonic speeds—far surpassing the speed of the Concorde, the British-French jetliner that flew for the first time in 1969—could be accommodated only in super-straight underground tunnels whose construction would claim all but a small share of the system's overall cost. By 1978 Salter was suggesting that the "Planetran" could be "extended to a worldwide network using under-ocean tunnels to connect continents" and that it would be "safe, convenient, low-cost, efficient and non-polluting." What a perfect example of that common phenomenon of an inventor attached to his cherished project far beyond the boundaries of any critical appraisal!

In reality, during the 1970s and 1980s the US saw only a further deterioration of its by then badly outdated railroad network even as Japan and Europe were expanding their high-speed links, starting with the Tokyo-Kyoto *shinkansen* in 1964 and the Paris-Lyon *Train à Grand Vitesse* in 1981. Concurrently, several countries, most notably Japan and Germany, constructed short tracks to begin experiments with magnetic levitation trains. Germany's Emsland track (1984–2012) was closed after a fatal accident, while Japanese researchers eventually (in 2015) achieved a new record speed of 603 km/h. The first short commercial maglev projects were the Pudong-Shanghai line in 2004, which used a German design, and Japan's Linimo line in 2005, with three more short and relatively low-speed connections coming online in South Korea and China in 2016 and 2017. Construction continues on the first long-distance maglev link, Japan's Chuo shinkansen between Tokyo and Osaka, but completion has been repeatedly postponed, now into the late 2020s.

There have been numerous bold intra- and international plans for maglevs outside East Asia, both in North America and in Europe, but no actual commitments. Publication of Hyperloop Alpha—greeted by the media and by new-tech enthusiasts unaware of the long history of kindred designs as amazingly original and stunningly transformative—brought a multitude of new plans for high-speed transportation links and led not only to a large number of naïve endorsements, numerous technical assessments, and exploratory designs but also to the setting up of new companies dedicated to turning the idea into commercial reality.

Virgin Hyperloop One, one of Richard Branson's companies, has by far the most ambitious plans: it operates a small 500-meter test track in

Nevada, and in 2020 its experimental pod with two passengers reached 175 km/h, hardly a remarkable achievement (higher speeds have been routine for high-speed trains since the 1960s). The Virgin Hyperloop has identified eleven possible routes in the US, including a megaproject linking Cheyenne in Wyoming (fewer than 600,00 people) with Houston (more than 1,800 kilometers away), nine routes in Europe, including undersea links between Corsica and Sardinia and between Spain and Morocco, and it has plans for lines in India (Pune to Mumbai), Saudi Arabia (Riyadh to Jedda), and the United Arab Emirates.

Hyperloop TT, a company with a 320-meter test track in France, has plans to connect such unlikely pairs of smaller (and relatively close) cities as Brno in Moravia with Bratislava in Slovakia and Vijaywada with Amarvati in India's Andhra Pradesh state. The earliest reports claimed completion of the first commercial hyperloop lines as soon as 2017, then 2019 and 2020. These years have come and gone, and we are no closer to even a convincing full-scale prototype demonstration, to say nothing about a single completed and truly reliable, safe, and profitable commercial link between two cities. No hyperloop line, on pylons or in tunnels, was in operation by early 2022, and the forecasts of earliest completion dates have shifted to the late 2020s. None of the system's often repeated advantages in comparison with high-speed rail—the absence of wheels (moving on air cushion or magnetically levitated), much faster operating speeds, significantly reduced energy use, lower construction costs—has been tested on even a single commercial project, and all such claims, until proven otherwise, remain in the category of wishful thinking.

Past proposals of rapid enclosed conveyances could not be realized because nineteenth- and twentieth-century engineers lacked suitable materials and techniques to build the requisite tubes and pods, to lower their inside pressure to levels approaching vacuum, and to propel the capsules safely and reliably across great distances. None of those challenges has gone away. Those who are best placed to appreciate the enormous difficulties facing such projects—vacuum physicists and railway engineers—have pointed out many fundamental barriers that would have to be overcome before vacuum tubes conveying people at near-sonic speeds could become even one-tenth as common as high-speed trains with steel wheels on steel rails.

Musk has trivialized many aspects of high-speed in-tube transportation, starting with route selection and ending with the actual cost of the entire system. To say that hundreds of kilometers of elevated tubes on pylons would "cause minimal disruption to farmland roughly comparable to a tree or telephone pole, which farmers deal with all the time" is a blatant misrepresentation of the need for actual pad sizes and access required for construction and maintenance. More important, as judged by numerous precedents in designing projects ranging from freeways to high-voltage transmission lines, route selection and approval would be a very complicated process marked by detours and delays.

Yet in July 2017 Musk, out of the blue, famously tweeted that he had "just received verbal government approval for The Boring Company to build an underground NY-Phil-Balt-DC Hyperloop. NY-DC in 20 mins." Anybody aware of the complex preparations, assessments, and negotiations leading to the approval of any multibillion-dollar project cutting across numerous jurisdictions and requiring the consent and cooperation of federal, state, and local governments, as well as compliance with a long array of applicable restrictions and requirements, must view Musk's 2017 tweet with utter disbelief. The tweet clearly implies that somebody in D.C. just picked up the phone and gave "verbal government approval" to a company that had no experience and no record of completed projects to build a 600-kilometer-long tunnel for a 1,000 km/h train.

Moreover, in a country that has not been able to upgrade the old rail line between New York and Washington, D.C., to anything better than Acela, a "rapid" train that does not have a dedicated track and whose average speed is just 125 km/h, even as more densely populated Europe has been building thousands of kilometers of special tracks for true rapid trains traveling 200–300 km/h and even more densely populated China has completed tens of thousands of kilometers of high-speed tracks. Similarly, anybody even fleetingly familiar with the initial cost estimates and eventual cost overruns that have accompanied most large-scale construction projects of the last generation must view the totals for capital costs of assorted proposed lines as nothing but uncertain guesses.

Although we have the advantages of having both advanced materials and propulsion and control systems of unprecedented power and complexity, affordable construction and routine and competitive operation of these new transportation schemes are not imminent. Challenges range from the acceptability of traveling in a claustrophobic pod hurtling through a metal pipe at the speed of sound (overcoming that is not as easy as projecting images of beautiful landscapes on the pod's walls!) to mastering and ensuring a large number of engineering firsts, with the pressure differential being the most obvious fundamental concern. Although the Hyperloop would not maintain a perfect vacuum, the pressure of 100 Pa comes close enough: jetliners flying in the upper atmosphere move through air whose pressure is more than 200 times higher, while the Hyperloop would operate under an equivalent of upper stratospheric (50 kilometers above sea level) pressure.

Catastrophic decompression is one of the worst possible scenarios in flight, and in the context of an extreme pressure difference it would be far more deadly in a long near-vacuum tube containing people-carrying pods. A steel tube on pylons would have to be engineered to maintain the thousandfold pressure difference between its inside and outside walls that threatens to crush it, and it would have to do so reliably along hundreds of kilometers of the track while also supporting the pressure generated by the rapidly moving pods and coping not just with overall thermal expansion along its course but with the differential thermal expansion between the tube's top and bottom, an occurrence particularly significant in hot climates. With a common temperature variance of 50°C (–10 to +40°C), the system would require numerous expansion joints, each required also to maintain a near vacuum.

Undergrounding the tube would eliminate most of these concerns, but to do that would require achieving an additional engineering first: constructing tunnels spanning hundreds, if not thousands, of kilometers, many of them in earthquake-prone regions. Although modern tunneling has become remarkably mechanized, costs remain high. The Gotthard Base Tunnel in Switzerland, at 57 kilometers the world's longest, cost about $10.5 billion (nearly $200 million per kilometer) and took nearly seventeen years to finish. And, all too obviously, an extensive network of near-vacuum tubes would pose an easy terrorist target, with relatively minor explosions causing catastrophic decompressions.

And in 2022 we got a comprehensive indication of what transportation experts think about the idea. A worldwide survey by the International

Maglev Board indicated that transportation experts have rejected the Hyperloop plan, mainly because they believe that it underestimates operational and safety complexity, along with costs (for both infrastructure and operations). All in all, not even a half-baked idea, and given these critical sentiments and the actual post-2013 accomplishments, it would seem prudent to advise the cognoscenti of rapid travel who are waiting for the fifth mode of transportation coming to their cities to watch their diet and exercise in order to remain in good health and achieve a long lifetime. If the lessons of promises and claims raised since 1810 by Medhurst, Goddard, Bachelet, and Salter are even remotely applicable to this latest round of infatuation with in-tube travel, then longevity is imperative. Even if everything goes better than we can imagine, it will take a long time indeed before the first fare-paying travelers enter the pods in Cheyenne, Brno, or Vijaywada, are accelerated to near-sonic speed, and arrive at the next station, hundreds of kilometers away, in just minutes. After more than two hundred years of such dreams, we are still waiting.

NITROGEN-FIXING CEREALS

Our understanding of the world and our well-being rest, to an insufficiently appreciated degree, on the scientific and engineering advances made between 1867 and 1914. Those decades saw the invention and commercialization of internal combustion engines, electricity generation and electric lights and motors, the inexpensive production of steel, the smelting of aluminum, the introduction of telephones, the first plastics, the first electronic devices, and a rapid expansion of wireless communication. We also came to understand the spread of infectious diseases and the nutritional requirements for healthy growth (above all, the need for adequate protein intake), as well as the need for indispensable plant nutrients in securing abundant and affordable food supply.

The latter realization was particularly important because the industrializing world was experiencing an unrepeatable combination of profound economic and social change. Rising demand for food and changing dietary habits, driven by faster population growth, mass-scale immigration to cities, higher disposable incomes, and rising female employment in factories and services, were key parts of this grand transformation.

Growing urban populations could afford not only to consume more plant foods per capita but also to buy more animal protein (meat, eggs, and dairy products) whose consumption was previously limited. Inevitably, this required diverting a growing share of harvests to animal feed, and the expanding mechanization of field tasks (tilling, seeding, harvesting) necessitated the maintenance of large numbers of draft animals: by the end of the nineteenth century, growing feed for America's horses and mules claimed about a fifth of the country's abundant farmland.

At the same time, average crop yields remained low (less than one ton per hectare for American and Russian wheat, no more than 1.5 t/ha even in the most productive European regions) even as the period of unprecedented farmland expansion (large-scale conversion of grasslands into cropland on North America's Great Plains and the Canadian prairies, as well as in Russia, South America, and Australia) was coming to its end. This combination of rising demand and limited prospects to meet it justified the quest for a fairly prompt solution—and thanks to advances in plant science, biochemistry, and agronomy, we understood, for the first time in history, what needs to be done to change this worrisome outlook.

The challenge and the solution were described in memorable terms in September 1898 by William Crookes, a chemist and a physicist, in his presidential address on wheat delivered at the British Association's annual meeting in Bristol. The most quoted sentence from his presentation was that "all civilised nations stand in deadly peril of not having enough to eat," and he estimated that the rising demand would bring a global wheat supply shortfall as soon as 1930. But he also identified the most effective solution and its most important component: increased crop fertilization and higher applications of nitrogen, the macronutrient that most often limits wheat (and indeed all cereal) yields. Crookes correctly observed that neither the animal manures nor the planting of green manures (alfalfa, clover) could meet future needs, and that the supply of the era's only important inorganic fertilizer, Chilean nitrates mined in the desert of Atacama, was obviously limited.

What was needed was to tap the unlimited supply of atmospheric nitrogen, to change the inert molecule (N_2) that forms nearly 80 percent of air's mass into a reactive compound (preferably ammonia, NH_3) that could be assimilated by crops and supply the macronutrient guaranteeing higher yields. As Crookes put it,

> The fixation of nitrogen is vital to the progress of civilised humanity. Other discoveries minister to our increased intellectual comfort, luxury, or convenience; they serve to make life easier, to hasten the acquisition of wealth, or to save time, health, or worry. The fixation of nitrogen is a question of the no far-distant future. . . . It is the chemist who must come to rescue. . . . It is through the laboratory that starvation may ultimately be turned into plenty.

Remarkably, that salvation appeared just a dozen years after Crooke's appeal. In 1909 Fritz Haber, a professor of chemistry at the University of Karlsruhe, succeeded in synthesizing ammonia from its elements (fig. 4.4). He did that by taking nitrogen from the air and hydrogen from reacting glowing coke with water vapor and combining the two elements under high pressure in the presence of a metal (iron) catalyst. His research

Figure 4.4 Fritz Haber (1868–1934), *left*, first demonstrated the synthesis of ammonia from its elements. Carl Bosch (1874–1940) turned the concept into engineering reality.

was supported by BASF, at that time the world's leader in the production of industrial chemicals, and it was under the leadership of Carl Bosch, one of BASF's most capable engineers, that Haber's bench-top demonstration was converted rapidly into a full-scale industrial synthesis (fig. 4.4).

BASF began synthesizing ammonia in September 1913, but the compound was soon diverted as a feedstock for the production of explosives during World War I, and fertilizer production resumed after 1918, but large-scale use of compounds derived from synthetic ammonia (urea, ammonium nitrates, and sulfates) had to wait until after World War II. Intensifying fertilization became a critical component of the Green Revolution that began to advance during the 1960s and relied on new short-stalked cultivars, high rates of nitrogen fertilization, and the application of pesticides to reach record cereal yields. By 1970 the global applications of synthetic nitrogen fertilizers were more than eight times the 1950 level. By the century's end they had risen above 80 million tons a year, and recently they have been close to 120 million tons of nitrogen a year.

Their benefits are indisputable: I have calculated that no less than 40 percent of the global population receive their dietary protein (directly from crops and indirectly from animal foodstuffs) from harvests that got nitrogen from the Haber-Bosch synthesis of ammonia; in China, the share is about 50 percent. Like nearly all beneficial inventions, however, this admirable solution has its drawbacks. To begin with, more than half of the applied nitrogen does not end up in crops but escapes through different routes (volatilization, leaching, erosion, bacterial conversion to nitrous oxide) into the environment. The global average for the share of nitrogen applications eventually ending up in harvested crops is now below 50 percent, and in China's intensive rice farming the share is only about a third.

During the second decade of the twenty-first century, worldwide applications of nitrogenous fertilizers averaged about 110 million tons a year, and losing half this mass is releasing more than 50 million tons of the element (in reactive compounds, mostly as nitrates and ammonia) into the environment. Moreover, the impact is highly concentrated in the Northern Hemisphere's agricultural regions where common annual applications average more than 100 kilograms of the nutrient per year per hectare, and for the most intensively cultivated corn or rice they surpass

200 kg N/ha. This is, of course, a substantial economic loss (nitrogen fertilizers commonly account for a fifth of variable expenses in intensive crop farming) and one that also causes major environmental problems.

None of these environmental problems is now more widespread and difficult to control than the creation of large dead zones in coastal waters. Nitrogen leached into streams is transported into ponds and lakes, eventually reaching the shallow coastal ocean waters, where it supports the excessive growth of algae. When these algae die and sink to the bottom, their decomposition consumes dissolved oxygen and leaves the water anoxic, suffocating fish and marine invertebrates. These dead zones are now found in the Gulf of Mexico and along many European and East Asian shorelines. Nitrogen oxide and nitrogen dioxide released from fertilization and converted to nitrates in atmospheric reactions contribute to acidified precipitation (the phenomenon popularly known as acid rain, whose generation is dominated by the emissions of sulfur oxides).

Another side effect of fertilization that is receiving more attention is the generation of nitrous oxide by bacterial decomposition of nitrates. Not only is N_2O a greenhouse gas but, on a hundred-year time scale, it has a nearly three hundred times higher global warming potential than carbon dioxide, the dominant greenhouse gas. But because of its relatively small emissions, N_2O is responsible for only about 6 percent of recent anthropogenic greenhouse gas emissions. Prolonged applications of heavy doses of synthetic nitrogenous fertilizers also affect a soil's natural fertility by leading to the decline of soil organic carbon (previously obtained from the recycling of manures and crop residues) and to diminished soil biodiversity. Reducing the fertilizer applications to the minima compatible with maintaining good yields is thus one of the foremost goals of modern agronomy.

But in contrast to nitrogen-hungry staple cereals, leguminous plants (seed-yielding peas, beans, lentils, soybeans, and peanuts, and cover crops, including clovers, vetches, and alfalfa) do not need any, or need only minimal, fertilizer applications not only to produce good yields but to leave residual nitrogen in soils after their harvest. This difference has been known since antiquity as traditional cultivators, ignorant of any macronutrient needs for crop cultivation, grew together leguminous and grain crops and rotated cereals with leguminous plants to improve the yields of cereals. But they had no idea why it was so, and it was only in 1838 that a French chemist, Jean-Baptiste Boussingault, demonstrated, by experimenting with peas grown in sterile sand, that legumes actually add nitrogen to the soil. The only way to explain this ability was to conclude that legumes can use the inert atmospheric nitrogen to produce reactive compounds, but the actual mechanism remained unknown.

It took another half century for two German chemists, Hermann Hellriegel and Hermann Wilfarth, to deduce, in 1888, that leguminous species are fundamentally different from grasses, be they wild varieties or cultivars of wheat, rice, barley, or oats selected for higher yields. Leguminous plants cannot themselves assimilate the free atmospheric nitrogen but obtain it through symbiosis with bacteria residing in their root nodules (fig. 4.5). Within a few years after this realization, microbiologists identified the bacteria residing in those nodules (belonging to the genus *Rhizobium*) as well other free-living bacterial fixers living in soils or in water.

What we have to do by using very high pressures and high temperatures—in large modern ammonia plants Haber-Bosch synthesis proceeds

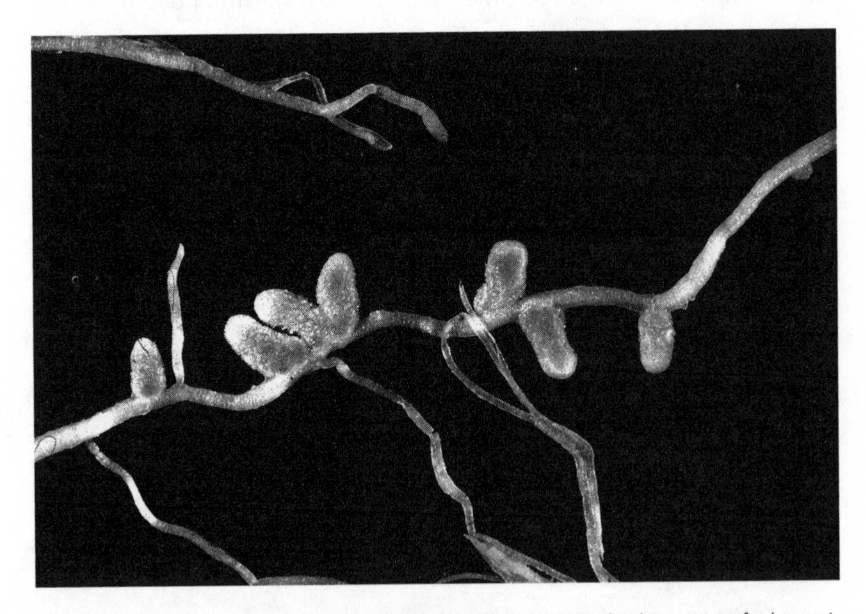

Figure 4.5 Nodules containing nitrogen-fixing bacteria attached to roots of a leguminous plant. *Source:* Matthew Crook. Reprinted with permission.

under a pressure that is 200–400 times that of atmospheric pressure at sea level and at temperatures higher than 400°C—*Rhizobium* bacteria can accomplish at ambient pressures and temperature, thanks to nitrogenase, an enzyme made up of two proteins (an FeMo and an Fe protein) that enable the reaction of hydrogen with nitrogen to form ammonia. But this nitrogen biofixation has a high energy cost, and nitrogenase does not tolerate oxygen and becomes irretrievably inactivated in air, a reality complicating its possible transfers.

This discovery of rhizobial nitrogen fixation suggested an intriguing possibility: might it be possible to induce cereals to behave as leguminous plants and fix all or most of the needed nitrogen through symbiosis with diazotrophs, the nitrogen-fixing bacteria, attached to their roots? Already in 1917 Thomas Burrill and Roy Hansen, researchers at the University of Illinois Agricultural Experimental Station, published a report titled "Is Symbiosis Possible between Legume Bacteria and Non-Legume Plants?" For decades this remained just a fascinating idea without any practical options for its gradual realization. But as our understanding of plant and bacterial physiology and genetics kept on advancing, the option, obviously still difficult to realize, seemed to become something that might be achievable in the not too distant future. Nobody voiced this hope more clearly than Norman Borlaug, an American agronomist, when receiving the Nobel Peace Prize in 1970 for his leadership in developing new, high-yielding crop varieties enabled by heavy nitrogen applications.

As he came to the conclusion of his Nobel lecture, Borlaug resorted to a dose of wishful science fiction, when he saw in his dream

> green, vigorous, high-yielding fields of wheat, rice, maize, sorghum and millet which are obtaining, free of expense, 100 kilograms of nitrogen per hectare from nodule-forming, nitrogen-fixing bacteria. These mutant strains of *Rhizobium cerealis* were developed in 1990 by a massive mutation breeding program with strains of *Rhizobium* obtained from roots of legumes and other nodule-bearing plants. This scientific discovery has revolutionized agricultural production for the hundreds of millions of humble farmers throughout the world, for they now receive much of the needed fertilizer for their crops directly from these little wondrous microbes that are taking nitrogen from the air and fixing it without cost in the roots of cereals, from which it is transformed into grain.

The benefits of such a symbiosis would be obvious. Cereal cropping would be more profitable thanks to much-reduced needs to buy and apply synthetic nitrogenous fertilizers. The environmental benefits of biofixation would range from much-reduced volatilization and leaching of applied fertilizers (hence less water pollution and acid precipitation) to lowered emissions of greenhouse gases and better soils (less compaction, more organic matter, higher nitrogen content). Research into extending legume-like nitrogen fixation to cereals began in earnest during the 1970s and has been pursued, with varying degrees of intensity, ever since.

The International Rice Research Institute had a project to assess opportunities for diazotrophs in rice, and similar projects have targeted other cereal crops in the US, Canada, the UK, and India, with financing provided by governments, universities, foundations, and companies. By the mid-1980s a major nitrogen fixation symposium ended by concluding that little of the experimental progress "has yet been applied in a practical sense to improve crop production"—but then the expectations of future success rose as genetic engineering produced some commercially very successful cultivars. The first insect-resistant varieties of corn and soybeans, incorporating insecticidal genes from *Bacillus thuringiensis*, were released in the US in 1996. Genetically modified canola (rapeseed grown for cooking oil) became available in 1995, and the US now also grows transgenic papaya, potato, alfalfa, sugar beets, and apples.

There are three distinct strategies to bring nitrogen fixation to cereals. The first and the most obvious approach was suggested more than a century ago on the discovery of symbiotic nodular biofixation: to replicate the arrangement common in legumes, to find the way for rhizobia and cereals to enter into the same kind of mutually beneficial interaction as rhizobia and leguminous plants have, to induce cereal crops to develop root nodules that would supply a significant portion of their nitrogen needs. Some plant scientists think that the best approach is to start by trying to reintroduce nitrogen fixation into non-nodulating, non-crop species that are evolutionarily more related to nodulating plants. In any case, this option should make every biologist think about Dobzhansky's often cited maxim that "nothing in biology makes sense except in the light of evolution."

And evolution (more than 100 million years of diversification among higher plant species) has not endowed a single nutritionally important species outside of the Leguminosae family with the capacity for symbiotic

Rhizobium-driven nitrogen fixation. This absence of symbiotic rhizobia outside the legume family (the only other notable nitrogen-fixing symbiosis, that of filamentous bacteria of the genus *Frankia*, affects some two hundred nonfood plant species) is even more remarkable insofar as nitrogen is the most common growth-limiting factor for all plant species, and yet evolution furnished only a small number of them with the means to alleviate this constraint.

Beyond this puzzle, there are other practical concerns. We know that legumes channel 10–20 percent of their energy (carbon) production to nodules, but because of enhanced photosynthetic capacity enabled by the higher nitrogen supply, this relatively high energy cost does not translate into a commensurate loss of yield. But this may not be the case with cereals endowed with symbiotic fixation capability. This means that nitrogen-fixing cereals might not find ready acceptance in those Asian nations and regions where the highest yields are needed to support large, high-density populations.

The second option for getting nonleguminous crops to fix nitrogen is to enhance the activities of bacteria that might be present in the root zone of cereal plants (free-living, not in rhizobia-like aggregations) in order to provide a more significant share of a crop's nitrogen demand or to introduce diazotrophs into the plant tissues of nonleguminous crops, either by treating their seed before planting or by using foliar sprays. This option became conceivable with the discovery of bacteria associated with tropical grasses. There are many free-living (nonsymbiotic) nitrogen-fixing bacteria, commonly including *Pseudomonas* and *Azospirillum* species in soils and cyanobacteria (*Nostoc, Anabaena* and many others) in waters, that thrive without any association with roots or other plant organs and that generally contribute only a modest amount of nitrogen to crops.

But during the late 1960s the Brazilian microbiologist Johanna Döbereiner led a research group that discovered several bacteria (*Acetobacter, Azospirillum, Herbaspirillum*) forming associations with the roots of some tropical grasses (fig. 4.6). These diazotrophs do not live in organized, visible root nodules that interact symbiotically with the host plant but are dispersed on and near plant roots, absorb some of their exudates, and indirectly transfer some of the nitrogen they fix. Later it was found that associative nitrogen fixation by *Azospirillum* living in the root zone

Figure 4.6 Johanna Döbereiner (1924–2000), Brazilian microbiologist and the pioneer of nitrogen fixation studies in corn and sugar cane. *Source:* Embrapa, Brazilian Ministry of Agriculture.

of cereals is sometimes a non-negligible contributor to the total nitrogen needs of rice and corn.

These discoveries opened up the possibility of enhancing the presence of associative bacteria living in the proximity of cereal roots, but any effective realization of this goal would require a much better understanding of the conditions that promote these associations and of the realistic uptake maxima we could expect in light of the low concentrations of nitrogen fixed by associated diazotrophs: even if successful, this effort would be of only very marginal help. A much better prospect emerged with the 1988 discovery of endophytic (living inside plant tissues) *Gluconoacetobacter diazotrophicus* diazotrophs in Brazilian sugar cane by Johanna Döbereiner and Vladimir Cavalcante. Later research found that *Herbaspirillum, Azoarcus,* and *Azospirillum* species are also involved. As a result, it remains difficult to distinguish the endophytic and nonendophytic contributions.

Azotic Technologies, a British company established by David Dent and Edward Cocking, is now offering a patented treatment using *Glucono-*

acetobacter diazotrophicus. Its initial product was a liquid seed inoculant, and now it also offers foliar treatment; it claims that these applications provide every cell in the plant with the ability to fix its own nitrogen and that the treatment's efficacy was proven on both corn and wheat in the UK, the US, Canada, Germany, Belgium, and France, and on rice in Vietnam, Thailand, and the Philippines. The company's US field experiments done with corn indicated average yield increases of 5–13 percent, in some cases even 20 percent in trials where nitrogen fertilizer levels had not been reduced; the mean response in Asian trials with rice was 17–20 percent.

On its website the company claims that the treatment can replace up to half the crop's nitrogen needs. In the US and Canada the treatment is marketed as Envita (either in-furrow or as foliar application) and promises a risk-free way to increase corn yields by at least 2.5 bushels per acre (that is nearly 160 kg/ha). But independent in-furrow trials done by Practical Farm Research in 2020 and 2021 in Kentucky, Illinois, Ohio, and Minnesota showed that some control plots (those receiving no Envita) actually yielded slightly more, and that the typical yield enhancement was no higher than a few percent. Similarly, Iowa foliar treatment trials showed that Envita had no significant effect on corn yields in five trials (with untreated plots having slightly higher yields), in one trial it provided a significant advantage, and in another one it resulted in a significant yield loss. Obviously, these results stand in a great contrast to advertised claims.

The third path, the most radical and most ambitious one, is having the receptivity to symbiosis encoded as a permanent plant trait, that is, to design new crops that could fix their nitrogen without any microbes thanks to the introduction of nitrogen-fixing (*nif*) genes directly into cereal plants. This is made difficult by two natural obstacles: the complexity of nitrogenase, the enzyme that is the essential catalyst needed to convert inert atmospheric nitrogen to ammonia (it also requires inputs of iron and of the much rarer molybdenum), and the enzyme's sensitivity to the presence of oxygen.

Canadian gene-transfer research has focused on triticale, a hybrid of wheat and rye, because the procedures work more efficiently in this crop than in wheat, and on transferring the entire cluster of *nif* genes using a nanocarrier (cell-penetrating peptides) into mitochondria. American researchers at MIT have been working with tobacco plants, a favorite for plant genetics experiments, to transfer *nif* genes from rhizobial bacteria. The task is very difficult not only because many genes are involved in the process but because gene expression and the cellular components directing the process are very different in bacteria and in plants. In 2018 progress was reported on assembling *nif* genes into a smaller number of "giant" genes that could be expressed as large proteins in host cells and then cut by special enzymes to release individual *nif* components.

The difficulties would not stop with having a genetically engineered cereal crop that actually fixes nitrogen, however: we must keep in mind past and recent travails with transgenic crops. Such crops have been welcomed by producers in North and South America (and generally accepted by consumers in those countries), but they have been shunned by nearly all EU countries and Japan, while China and India grow transgenic cotton but no genetically modified staple food or animal feed crops. This reluctance or outright rejection rests on widespread public fears that will not easily be tempered. Transgenic crops have run into opposition from vocal green and organic lobbies arguing against any genetic modification of foodstuffs.

And it is one thing to genetically modify grain corn for animal feeding and another to tinker with wheat, the staple of nutrition and one of the foundations of Western civilization. As a result, genetically modified wheat varieties have been developed and tested, but none is commercially produced in North America, Europe, Asia, or Australia. In the US, Canadian, and Australian cases there is an additional obvious concern: these countries are major grain exporters and would not be able to send their wheat to most of the world's countries that do not accept any genetically modified food. In October 2020 the Argentinian Ministry of Agriculture approved a transgenic drought-tolerant wheat cultivar, Bioceres HB4, for human consumption: is this the beginning of a trend or an inconsequential exception? And where do we stand more than 130 years after the elucidation of symbiotic rhizobial fixation, more than a century after Burrill and Hansen asked whether the symbiosis between diazotrophs and cereals was possible, fifty years after Borlaug had his Nobelian vision, and after decades of intensifying advances in genetic engineering?

Real breakthroughs were promised during the 1970s and even more optimistically during the 1990s. Will the 2020s, thanks to the advances in genetic engineering, be able to deliver? As might be expected with modern media reporting, every news report of some notable research advance has been commonly seen as moving us "closer" to the holy grail of nitrogen fixation in cereals—but "closer" remains elusive. "Substantial progress" reported in one year has no consequences five years later. Some claims play loose with timing. On its website, Joyn Bio, a new joint venture between Gingko Bioworks (a Boston outfit that creates custom-made bacteria) and Leaps by Bayer (now, well beyond aspirin, a leading agribusiness company), says "our first product *is* an engineered microbe that enables cereal crops like corn, wheat, and rice to convert nitrogen from air," but scrolling down for details one finds "our first product *will be* an engineered microbe that enables. . . ." Emphasis is mine.

Giles Oldroyd, at the Crop Science Center at Cambridge University, offers the best and the only honest reply to the question about how long it will take to get nitrogen-fixing cereals: "There is no answer to that. We are working in the unknown." The contours of this unknown have, inevitably, shrunk thanks to the decades of agronomic, plant science, and genetic research, but even so we cannot claim that highly rewarding plantings of soybean-like wheat or lentil-like rice are coming to nearby fields by a certain date, guaranteed to maintain yields with much-reduced nitrogen fertilizer applications and with the most welcome side effects of multiple environmental benefits.

CONTROLLED NUCLEAR FUSION

Size- and radiation-wise, there is absolutely nothing remarkable about the star at the center of our planetary system: millions of very similar stars can be found among some 100 billion radiant bodies that fill our galaxy. Because of its characteristic yellow color, astronomers put it just about in the middle of a diagram that classifies stars by the spectrum of their light. Sizewise, it is a very common astral dwarf of the G2 V class, as is Proxima Centauri, the closest star outside our solar system. The early Sun radiated nearly a third less energy than the star does now, about 4.5 billion years after its formation. Its ordinariness may be universally unremarkable but its energy production is astonishing: the Sun's luminosity is about 3.8 $\times 10^{26}$ watts (joules per second), while the world's total primary energy consumption (all fuels and all hydro, nuclear, wind, and solar electricity) is about 1.8×10^{13} watts, a difference of thirteen orders (tens of trillions).

How it is done must have puzzled many observers in the premodern past, but only the nineteenth century began to provide analytical tools that could be used to explain that extraordinary outpouring of energy. The most obvious analogy with a common terrestrial process would be combustion, but that transformation (chemically, a rapid oxidation) could not release enough energy to generate the immense astral heat and light (burning a gram of carbon releases 30 joules, burning a gram of hydrogen releases 113 joules). In a paper published in 1848, Robert Mayer, a German physician and physicist and a founder of thermodynamics, concluded (wrongly) that the Sun's heat arises from the energy of meteorites that are falling into it, and in 1854 Hermann Helmholtz, another German physicist, suggested that the Sun could generate enough energy by converting gravitational motion to heat: attracted by the force of gravity, the Sun's outer layers could be moving inward, making a slowly shrinking star bright and very hot. An annual contraction of a mere 40 meters—at the Sun's diameter (1.393 million kilometers), clearly too small to be detected across millennia of human observations—would be enough to produce the amount of energy that it was radiating in the mid-nineteenth century and would have sufficed to produce it for some 10^{15} seconds, or more than 30 million years.

This presented an obvious time constraint because the contemporaneous geological and biological studies indicated that the spans of terrestrial and organismal evolution had to be much longer than that. Was Darwin's insistence on long evolutionary life spans untenable in light of the claims of the physicists (something that indeed discomfited Darwin until his death), or did the physicists offer a wrong explanation? Once the studies of radioactivity (beginning in 1896 with Henri Becquerel's discovery of the phenomenon) put the Sun's age at about five billion years, it became clear that the gravitational hypothesis was untenable, and the search was on for a reaction that could be sustained across this enormous span of time.

In the1920s the British astrophysicist Arthur Eddington hypothesized that star energy comes from nuclear fusion and proton-electron annihilation and insisted that star interiors provided an environment hot enough to allow for such reactions. Finally, during the 1930s advances in nuclear physics made it clear that nuclear reactions drive solar radiation, and by the end of the decade it became clear how they proceed. The simplest possible sequence begins with the fusion of two protons to form heavy hydrogen (deuterium) and was suggested first by Carl Friedrich von Weizsäcker in 1937 and properly quantified by Charles Critchfield and Hans Bethe soon afterward. This reaction also produces a positron and a neutrino, and the deuterium fuses with another proton to produce an isotope of helium and releases an order of magnitude more energy than the first reaction.

Bethe also explained the second set of reactions, which begins with the fusion of carbon and hydrogen to produce an isotope of nitrogen and gamma radiation and ends with an isotope of nitrogen and hydrogen producing carbon and helium, and this earned him the Nobel Prize in Physics in 1967 (fig. 4.7). Carbon is used only as a catalyst; the reaction combines four protons and two electrons to form one helium nucleus,

Figure 4.7 Hans Bethe (1906–2005) received the Nobel Prize in Physics for explaining the Sun's fusion reactions. *Source:* Los Alamos National Laboratory.

and in 1938 Bethe found that "the carbon-nitrogen cycle gives about the correct energy production in the sun." The fusion of hydrogen into helium in the proton-proton cycle takes place only once the temperature reaches 13 million degrees of absolute temperature, and the regenerative carbon-nitrogen cycle comes to dominate the total rate above 16 million degrees.

Reactions in the Sun's core, proceeding under a pressure about 250 billion times more than at Earth's surface, consume 4.3 million tons of matter every second and release 3.89×10^{26} joules. This energy flux is rapidly converted into heat and transported outward, with every square meter of the Sun's visible light-emitting layer radiating about 64 MW. Very little of that flux is absorbed before it reaches Earth's orbit, and hence the power flux input available at the top of Earth's atmosphere, the solar constant, is nearly 1,370 W/m^2. These explanations were needed because the quest for controlled nuclear fusion is attempting nothing less than to replicate the extreme circumstances that sustain the Sun's enormous energy output and then use the resulting heat to generate electricity.

Replicating the Sun's nuclear fusion in the format of a one-time and virtually instant energy release—that is, using nuclear fusion as the source of unprecedented explosive power—was accomplished in just fourteen years after Bethe's 1938 solution. In 1938, he and all other prominent American physicists (led by Robert Oppenheimer and including Ernest Lawrence, Glenn Seaborg and Philip Abelson) and expatriates who had left Europe for the US (most prominently Enrico Fermi, Leo Szilard, John von Neumann, and Edward Teller) could not know that in just a few years they would be cooperating on designing and building the world's first nuclear weapons using the nuclear fission of the heavy elements. The Manhattan Project's work began in earnest in 1942, Bethe became the head of its theoretical division, the first fission weapon was tested in July 1945, and Hiroshima and Nagasaki were bombed on August 6 and 9, 1945, respectively.

By that time Edward Teller and Enrico Fermi were already thinking not only about a fusion bomb but also about the possibility of controlled fusion reactions, and six years later, Teller's solution to the critical physics design problem led to the first American test of a hydrogen bomb (actually a stationary device weighing seventy-four tons) in 1952 (fig. 4.8).

Figure 4.8 Edward Teller (1908–2003) and the first US hydrogen bomb test. *Sources:* Lawrence Livermore National Laboratory; Comprehensive Nuclear-Test-Ban Treaty Organization.

Soviet physicists tested their first thermonuclear weapon (equivalent to 1.6 million tons of TNT) in November 1955. Both countries then went on building more powerful weapons, but while the US effort stopped after testing a 15 Mt-equivalent design in 1954, the Soviets actually tested a 58 million ton-equivalent bomb in October 1961. Less than two decades after Bethe's explanations, it became possible to replicate the fusion powering the stars in weapons of unprecedented power and energies capable of instantly obliterating even multimillion-person cities.

As we have already seen, commercial deployment of US nuclear fission relied heavily on the previously developed military application and on the pressurized water reactor designed to power submarines. There was no parallel in the quest for commercial fusion: a hydrogen bomb could not be repurposed for generating useful industrial or space heat or to generate electricity. What was required was a device that would keep plasma confined long enough to initiate nuclear fusion. The easiest (a relative term in this context) way to achieve controlled fusion is to combine the two heavy isotopes of hydrogen, deuterium and tritium, to form an isotope of helium.

At the temperatures required for nuclear fusion, hydrogen isotopes exist in the form of plasma, a superheated state of matter with electrons stripped away from nuclei and forming ionized gas. Enormous energy is needed to fuse the nuclei of deuterium and tritium: they must have a mutual kinetic energy of at least 100 keV (100,000 electron volts). With 1 eV equivalent to 11,606 K (degrees of absolute temperature), this is equivalent to roughly 110 million degrees. First, two deuterium atoms release a proton and a triton: $^{2}H + {}^{2}H \rightarrow {}^{3}H + {}^{1}H$. Then another deuteron and the triton fuse at a mutually high kinetic energy and produce helium-4 (an energetic alpha particle) and an even more energetic neutron: $^{2}H + {}^{3}H \rightarrow {}^{4}He + {}^{1}n$. The helium carries 20 percent of all energy produced by this fusion, and once it provides enough energy, plasma becomes hot enough to be maintained without any need for external energy input (the self-heating "burning" plasma stage).

Neutrons carrying 80 percent of the energy produced by fusion escape from the plasma and their subsequent absorption elsewhere generates the heat that would eventually be used to generate electricity in the same way it is produced in large fossil-fueled stations (by using steam turbo generators). The two elements needed for controlled fusion are abundant: deuterium can be separated from ocean water (there are 33 grams of deuterium in every cubic meter of seawater) and lithium. This light metal is now in high demand for batteries, but its resources (close to 90 million tons in 2020, and highly likely to grow further) are sufficient for about one thousand years of extraction at the recent level. But future fusion plants will also need to generate their own tritium because this isotope is exceedingly rare in nature (it is produced in the atmosphere through the collisions of cosmic rays with nitrogen molecules). This tritium generation would be done by capturing neutrons in a lithium blanket surrounding the confined plasma.

Three conditions must be met to enable the highly energetic collisions of nuclei that make fusion possible: maintaining extraordinarily high temperatures, sustaining a plasma density high enough to increase the probability of nuclear collisions, and providing sufficiently long-lasting plasma maintenance required for continuous heat generation. These are high barriers to overcome, but when the quest for controlled fusion was in its beginnings, many physicists felt that relatively speedy success was possible. Homi Bhabha, the Indian nuclear physicist who chaired the First International Conference on the Peaceful Uses of Atomic Energy in Geneva in 1955, said during his opening speech that "a method of controlled release of the energy of nuclear fusion would be discovered in the next 20 years."

This was just the first of many exaggerated expectations, but the two decades between 1955 and 1975 saw some notable experimental progress, notably at the Kurchatov Institute in the USSR, where research on controlled fusion began in 1951.

Many Soviet physicists believed that magnetic confinement would offer the best path to eventual success. This could be done in a number of complex configurations, but a toroidal (doughnut-like) device, with plasma kept coursing within a tubular vacuum chamber by means of extraordinarily powerful magnets, was their preferred design. The concept, patented in 1946 by M. Blackman and G. P. Thomson in Britain, was developed in the early 1950s with contributions by Igor Tamm and Andrei Sakharov, and the design of this experimental magnetic thermonuclear reactor (with the first experiments done during the late 1950s) became universally known as tokamak, an acronym created from the beginning syllables (and the first letter) of the Russian term for toroidal chamber with magnetic coils, ***to**roidal'naya **ka**mera s **ma**gnitnymi **k**atushkami.*

Other notable experimental designs introduced during the 1950s were Lyman Spitzer's stellarator (using external coils to generate twisting magnetic fields to control the plasma) and the Z-pinch (relying on plasma compression by magnetic fields). Progress in achieving higher temperatures was very slow during the 1950s and 1960s, but by 1975 the best Soviet tokamak could produce a plasma temperature of 1 keV (or 11.6 million °K), and just three years later it reached 8 keV (92.8 million °K). Research on controlled fusion turned into a growth industry. In the US, participating institutions have included large government laboratories (Los Alamos, Lawrence Livermore, Oak Ridge, Lawrence Berkeley, Sandia, Savannah), universities (MIT, Columbia, Princeton, the University of California, Texas), and private companies, but the country accounts for only about a sixth of the global magnetic fusion effort, with most monies coming from the EU (about 40 percent) and Japan (about 20 percent). Since 1970 about sixty large-scale conceptual controlled-fusion designs were developed, and more than one hundred experimental facilities were built in the US, Russia, Japan, and the EU.

These efforts have attracted steady low-level media attention, which tends to spike with every new announcement of experimental progress. The attraction is understandable. Fusion has been described as the ultimate clean energy source, inexhaustible, sustainable, and carbon-free; a source that could be deployed anywhere and at any time: nothing less than the Sun in a bottle at our command. Yet another positive attribute came to the fore in the context of the recent preoccupation with greenhouse gas emissions: fusion would not, obviously, emit any CO_2 or CH_4. Unfortunately, the mass media have chronically misinterpreted all announcements of experimental controlled fusion advances in two ways. First, they have routinely labeled every gain as a breakthrough, and second and more important, they do not make it clear that these "breakthroughs" bring controlled nuclear fusion closer only in the sense of being a proven (rather than a theoretical) possibility, not to being an actual commercial operation widely deployable to generate heat and electricity.

Diligent searches would uncover scores of such reports; here are just a few English-language examples from the past four decades. *Science* in 1978: "Report of Fusion Breakthrough Proves to Be a Media Event." *Science Digest* in 1985: "Breakthrough Brings Fusion Closer." *Science* in 1989: "Fusion Breakthrough?" *The Hill* (magazine of the US Congress) in 2012: "Fusion Breakthrough Dawns a New Era for US Energy and Industry." *Science* in 2015: "Secretive Fusion Company Claims Reactor Breakthrough." *Clarion News* in 2021: "The Fusion Age Is upon Us." The MIT Office of Sustainability in 2021: "The Sun in a Bottle."

This Sun-in-a-bottle news release reported the demonstration of the world's strongest high-temperature conducting magnet, built by researchers at MIT and its spinoff, Commonwealth Fusion Systems. As is nearly always the case with these kinds of advances, it is impossible just to state the latest achievement; the news must be framed as "a breakthrough that paves the way for carbon-free power" putting us "a step closer towards a workable fusion reactor" and providing "reason for hope that in the not-too-distant future, we could have an entirely new technology to deploy in the race to transform the global energy system and slow climate change." Some comments were even more effusive: Dennis Whyte, director of MIT's Plasma Science and Fusion Center, claimed that "this magnet will change the trajectory of both fusion science and energy, and we think eventually the world's energy landscape."

Another 2021 fusion record was set at the US National Ignition Facility (at the Lawrence Livermore National Laboratory) in which light from massive lasers was concentrated on a tiny target, producing a miniature hotspot that generated more than 10 quadrillion watts of fusion power, though for a mere 100 trillionths of a second and amounting to only about the inputted laser energy. In August 2021 an experiment showed a twenty-five-fold increase over the previous (2018) record yield, putting the researchers "at [the] threshold of fusion ignition." This research, relying on inertial confinement (a capsule filled with deuterium and tritium is irradiated by lasers, x-rays, or particle beams to compress the fuel and heat it to ignition point) seemed to offer a possible alternative to magnetic confinement designs, but its progress has been slow.

Setting the media labels and overenthusiastic promotions aside, how close have we come? That question is commonly answered by noting how close we have approached Q, the ratio of the thermal power produced by deuterium-tritium fusion to the power injected into a fusion device in order to superheat the plasma and initiate a fusion reaction at usefully high levels. Obviously, Q = 1 is the breakeven point, and the history of fusion reactor designs can be traced by their rising Q factors. So far, the highest Q, 0.67, has been achieved by the European tokamak JET in the UK. The record-size tokamak, the International Thermonuclear Experimental Reactor (ITER), is now under construction in France and should not only reach the breakeven point but surpass it considerably. ITER's origins go back to a 1985 Geneva summit; the agreement to build it was signed in 2006, and construction began in Cadarache in southern France in 2010. The joint effort is supported by thirty-five countries, including the EU, Switzerland, Japan, Russia, the US, China, India, and South Korea (fig. 4.9).

Like every tokamak, ITER has central solenoid coils, large toroidal and poloidal magnets (respectively around and along the doughnut shape).

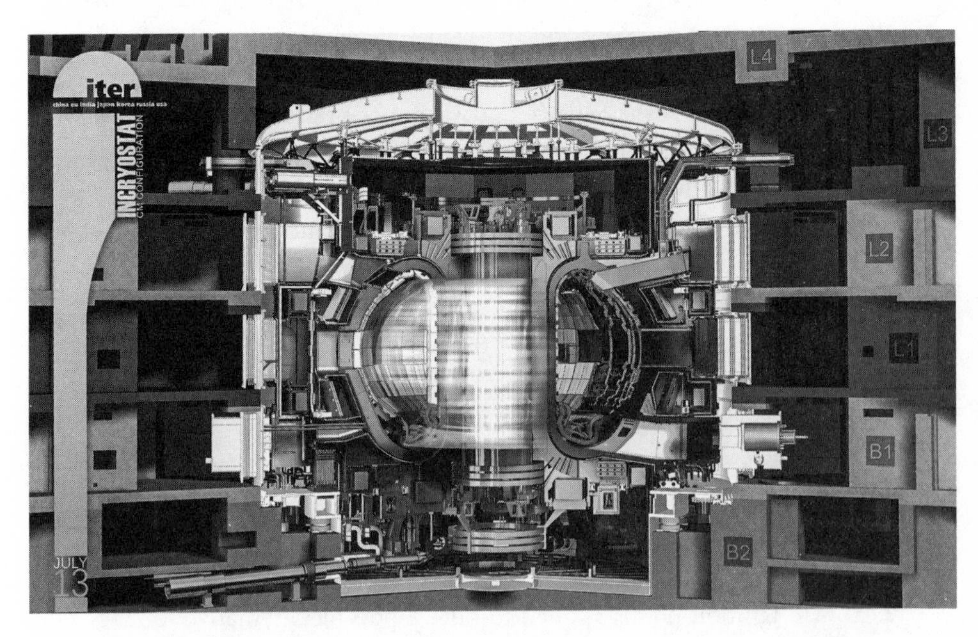

Figure 4.9 When completed (the projected dates of completion keep changing), the ITER tokamak will be the largest fusion device. *Source:* Image © ITER Organization, http://www.iter.org.

The basic specifications are a vacuum vessel plasma of 6.2 meter radius and 830 cubic meters in volume, with a confining magnetic field of 5.3 tesla and a rated fusion power of 500 MW (thermal). This heat output would correspond to $Q \geq 10$ (it would require the injection of 50 MW to heat the hydrogen plasma to about 150 million degrees) and hence would achieve, for the first time on Earth, a burning plasma of the kind required for any continuously operating fusion reactor. ITER would generate burning plasmas during pulses lasting 400 to 600 seconds, time spans sufficient to demonstrate the feasibility of building an actual electricity-generating fusion power plant. But it is imperative to understand that ITER is an experimental device designed to demonstrate the feasibility of net energy generation and to provide the foundation for larger, and eventually commercial, fusion designs, not to be a prototype of an actual energy-generating device.

ITER's operation was originally scheduled to start in 2016. By 2012 the time for the first plasma generation had been postponed to 2020, and in June 2016 the first plasma date was shifted to December 2025 (and the first operation with tritium to 2032). No matter when it becomes fully functional, ITER will not capture any outgoing heat to be used for electricity generation and will not attain a state of continuous fusion: it will generate pulsed net energy ($Q > 1$) only when the ratio is calculated by dividing the heat energy output by the energy used to heat the plasma (50 MW), not by the total electricity consumption of the facility. ITER's total electrical power demand will be about 300 MW, mostly for the cryogenic plant needed to cool the superconductor magnets to –269°C and power them to produce a 15 megaamperes plasma current.

Heat released by controlled fusion has always been intended to generate electricity: a conventional heat-exchange system would remove the heat from appropriate portions of the fusion reactor, and water would be then heated to produce steam for large generators in central plants. This roughly 140-year-old conversion has become steadily more efficient, with the latest turbo generators, operating under supercritical steam pressure, turning more than 40 percent of the incoming high-temperature heat into electricity. Consequently, if a fully functioning ITER were a real electricity-generating plant it would, with 40 percent efficiency, convert 500 thermal MW into 200 MW of electricity, for a net power loss of 100

MW. This means that any commercial fusion plant will have to operate with a substantially higher Q in order to produce electricity whose cost would be competitive with today's, and the future's, alternatives.

And when might that be? ITER is intended to be run for twenty years before construction is started on DEMO, a fusion demonstration power plant whose anticipated dates slip with ITER delays. The original 2021 ITER timeline had the DEMO operating in the early 2040s, but in 2017 it was announced that 2054 is an optimistic date. There are other, national timelines, ranging from India's and South Korea's DEMO construction starts in 2037 to Russian and American goals to have a DEMO-FNS operating by 2050 or shortly afterward, and if this accelerated development actually takes place, then one source claims that about 1 percent of global energy demand could be supplied by fusion by 2060.

These projections beyond ITER reflect the combination of ever-present technical uncertainties and funding commitments, and the typical estimates of an additional thirty to thirty-five years required before fusion delivers any practical benefit remain consistent with the estimates of a thirty-year success horizon that have been around since the 1950s. During the past seven decades the world has spent at least $60 billion (in 2020 monies) on developing controlled fusion, but it remains perhaps the most stubbornly receding *fata morgana* on record: always to be reached after yet another thirty years. Whenever I see a new recital of these always-distant dates, I remember what the late David Rose, MIT's physics professor who spent his professional life working on plasma physics and whom I came to know in the early 1980s, told me about the commercialization of fusion electricity generation: it might be at least as difficult, and possibly even more taxing, than getting to $Q > 1$.

The two key challenges are problems with containment and fuel assemblies and the large requirements for parasitic power. Neutron streams generated by deuterium-tritium fusion will damage the solid containment vessel by swelling, fracturing, and embrittlement, imperiling its integrity. Although the radioactivity level per unit mass of waste would be much smaller than that generated by fission reactors, the total mass of wastes would be many times larger and would require permanent off-site disposal. And unlike in the case of fission reactors, a significant share of this material damage and waste would be incurred not by producing any useful power. Fusion's parasitic power drain is due to the need to run liquid helium refrigerators, water and vacuum pumps, tritium processing, and other plant needs and to control the magnetic confinement: a small-size fusion plant rated at 300 MW and producing (with 40 percent efficiency) 120 MW of electricity would barely supply its continuous on-site needs, and only much larger projects, where the parasitic power would become a much smaller share of the overall rated output, could be economical.

And we can only guess at how much this effort will cost. Initial cost estimates for ITER were as low as €5 billion, but by 2016 ITER's director had admitted that the project was a decade late and at least €4 billion over budget, with later reports showing the overall sum reaching €15 billion, and in 2018 the US Department of Energy nearly tripled its cost estimate for ITER, to $65 billion. ITER leaders dismissed that claim, but by 2021 they were admitting to further, COVID-19-related delays and cost overruns. Even if the learning process is taken into account, the most optimistic estimate for demonstration reactors (at least three of them, to be built after 2040) would not be less than $20 billion each.

By the time we got to the beginning of commercial electricity generation, the cumulative 1950–2050 outlays on controlled fusion might be thus on the order of $200 billion. If the demonstration plants worked as promised, it would not be a bad bargain: $200 billion is roughly equal to the annual GDP of Greece, and less than President Biden budgeted in his 2021 infrastructure plan to boost the country's R&D capacity. And, to put it all into the most realistic perspective, it is just a tenth of the monies spent on the two-decades-long war in Afghanistan that ended in the chaotic US withdrawal and the Taliban's complete victory.

None of the delays, problems, or costs I have noted make any difference as far as the true fusion believers are concerned. The second pandemic year actually brought some of the most optimistic claims. A former undersecretary for science at the US Department of Energy wrote in September 2021 that "everyone should absorb that the fusion age is upon us. The target for net-energy-out fusion is now four years; not 30 years," and in October 2021 the *New Yorker* ran a long story on the "green dream" subtitled "Is Limitless Clean Energy Finally Approaching?" Early in 2022 came the announcement by a group of scientists led by the physicists at the Lawrence Livermore National Laboratory that they had used inertial fusion implosion to generate in the laboratory a burning plasma state during which brief duration the plasma was, unlike in all previous experiments, predominantly self-heated. This is a significant advance that puts the laser-initiated fusion closer to reality, but still not closer to any imminent commercial applications.

This brief account of controlled fusion efforts would not be complete without going back to 1989. Early in that year came (in a press conference and in a brief paper in the *Journal of Electroanalytical Chemistry*) a radical departure from the decades of news concerning advances in the quest for controlled thermonuclear power. Two physicists at the University of Utah, Stanley Pons and Martin Fleischmann, claimed they had succeeded in fusing deuterium nuclei at room temperature in a test tube. Electrolysis of a lithium salt solution led so many deuterium atoms to absorb into a palladium electrode that some of their nuclei appeared to fuse, producing net energy (above that supplied for electrolysis), as well as neutron and gamma ray emissions, clear signs of a process previously attainable only under starlike conditions, and proof of what the press soon called cold fusion.

There is no need to recapitulate the ensuing media frenzy and the extensive experimental efforts to replicate that sensational finding. Before the year's end an expert panel advised the Department of Energy not to fund further research. The entire field of cold fusion is now known as low-energy nuclear reaction (LENR), and the recent (2019) multi-institutional evaluation (motivated by the possibility that the past dismissal could have been premature) can be summarized by its conclusion: "Here we describe our efforts, which have yet to yield any evidence of such an effect." The evaluation's by-product was to acknowledge that "there remains much interesting science to be done in this underexplored parameter space"—but that could be said about countless other topics in modern science. But an Advanced Research Projects Agency's meeting in October 2021 heard a more upbeat presentation, which concluded that "LENR occur and they do, indeed, involve nuclear reactions," and that the experimental results "suggest practical promise."

LENR adherents have been publishing scores of new papers every year, and many advocates maintain that theirs is not a case of pathological science (Philip Ball's characterization in *Nature*) and that they will eventually be vindicated by giving humanity an endless source of energy with nothing more than some simple systems of heavy water and palladium electrodes. But the fact remains that after thirty-plus year of these claims, convincing proof is still missing, and even if it were found, the lessons of hot fusion would justify extreme skepticism before providing the dates of eventual commercial exploitation. Hot or cold fusion—we are still waiting.

5

TECHNO-OPTIMISM, EXAGGERATIONS, AND REALISTIC EXPECTATIONS

This book has only modest goals: to remind us that success is only one of the outcomes of our ceaseless quest for invention; that failure can follow initial acceptance; that the bold dreams of market dominance may remain unrealized; and that even after generations of (sometimes intensifying) efforts, we may not be any closer to the commercial applications first envisaged decades ago. And what is true about the past is, despite recent claims to the contrary, likely to be repeated in the future. This closing chapter provides some factual correctives to visions of the ever-faster progress of invention that are now detailed in best-selling books, and will deconstruct hyperbolic claims that now accompany so many announcements of recent advances achieved during this supposedly unrivaled era of innovation. This cautionary attitude should be self-evident to any diligent student of modern technical advances—and so should be the basic attendant lessons.

First, every major, far-reaching advance carries its own inherent concerns, if not some frankly undesirable consequences, whether immediately appreciated or apparent only much later: leaded gasoline, a known danger from the very start, and chlorofluorocarbons, found undesirable only decades after their commercial introduction, epitomize this spectrum of worries. Second, rushing to secure commercial primacy or deploying the most convenient but clearly not the best possible technique may not be the long-term prescription for success, a fact that was clearly demonstrated by the history of "beaching" the submarine reactor for a rapid start of commercial electricity generation.

Third, we cannot judge the ultimate acceptance, societal fit, and commercial success of a specific invention during the early stages of its development and commercial adoption, and much less so as long as it remains, even after its public launch, to a large extent in experimental or trial stages: the suddenly truncated deployments of airships and supersonic airplanes made that clear. Fourth, skepticism is appropriate whenever the problem is so extraordinarily challenging that even the combination of perseverance and plentiful financing is no guarantee of success after decades of trying: there can be no better illustration of this than the quest for controlled fusion.

But both the acknowledgments of reality and the willingness to learn, even modestly, from past failures and cautionary experience seem to find less and less acceptance in modern societies where masses of scientifically illiterate, and often surprisingly innumerate, citizens are exposed daily not just to overenthusiastically shared reports of potential breakthroughs but often to vastly exaggerated claims regarding new inventions. Worst of all, news media often serve up patently false promises as soon-to-come, fundamental, or, as the current parlance has it, "disruptive" shifts that will "transform" modern societies. Characterizing this state of affairs as living in a postfactual society is, unfortunately, not much of an exaggeration.

BREAKTHROUGHS THAT ARE NOT

In light of how common this category of misinformation concerning breakthrough inventions (and their likely speed of development and the ensuing impact on the society) has become, any systematic review of this dubious genre would be both too long and too tedious. Instead, I will first note the breadth of these claims—with impossible timings and details coming across the vast range of scales, from colonizing planets to accessing our thoughts—and then review in some detail their specific instances in three prominent areas abounding in alleged breakthroughs, those of drug discovery, aviation, and artificial intelligence.

In 2017 we were told that the first mission to colonize Mars would blast off in 2022, to be followed soon by an extensive effort to "terraform" the planet (turn it into a habitable world by creating an atmosphere) preparatory to its large-scale colonization by humans. As science fiction this was an old and utterly unoriginal fable: many storytellers have done that, no one more imaginatively than Ray Bradbury in his *Martian Chronicles* in

1950. As a prediction and description of an actual scientific and technical advance it is a complete fairy tale, but one that has been reported seriously and repeatedly by the mass media for years as if it were something that would actually get under way according to that delusional schedule.

At the opposite site of this touted invention spectrum (from transforming the planets to reconnecting individual neurons) is a way for machines to merge with humans' brains: the brain-computer interface (BCI) has been a much-researched topic during the past two decades. This is something that would eventually require the implanting of miniature electronic devices directly into the brain to target specific groups of neurons (a noninvasive sensor on or near the head could never be so powerful or precise), an undertaking with many obvious ethical and physical perils and downsides. But one would never know this from reading the gushing media reports on advances in BCI.

This is not my impression but the conclusion of a detailed examination of nearly four thousand news items on BCI published between 2010 and 2017. The verdict is clear: not only was the media reporting overwhelmingly favorable, it was heavily preoccupied with unrealistic speculations that tended to exaggerate greatly the potential of BCI ("the stuff of biblical miracles," "prospective uses are endless"). Moreover, a quarter of all news reports made claims that were extreme and highly improbable (from "lying on a beach on the east coast of Brazil, controlling a robotic device roving on the surface of Mars" to "achieving immortality in a matter of decades") while failing to address the inherent risks and ethical problems.

In light of these planet-molding claims and brain-merging promises, how much easier it is, then, to believe many comparatively down-to-earth achievements that have been wholesaled by media during recent years. Forecasts of completely autonomous road vehicles were made repeatedly during the 2010s: completely self-driving cars were to be everywhere by 2020, allowing the operator to read or sleep during a commute in a personal vehicle. All internal combustion engines currently on the road were to be replaced by electric vehicles by 2025: this forecast was made and again widely reported as a nearly accomplished fact in 2017. A reality check: in 2022 there were no fully self-driving cars; fewer than 2 percent of the world's 1.4 billion motor vehicles on the road were electric, but they were not "green," as the electricity required for their operation came mostly from burning fossil fuels: in 2022 about 60 percent of all electricity in general came from burning coal and natural gas.

By now, artificial intelligence (AI) should have taken over all medical diagnoses: after all, computers had already beaten not only the world's best chess player but even the best Go master, so how much more difficult could it be for the likes of IBM's Watson to do away with all radiologists? We know the answer: in January 2022 IBM announced that it was selling Watson and exiting health care. Apparently, doctors still matter! And the problems with electronic medicine affect even the simplest of tasks, the adoption of electronic health records (EHR) in place of charts written in longhand. According to a 2018 survey by Stanford Medicine researchers, 74 percent of responding physicians said that using an EHR system increased their workload and, even more important, 69 percent claimed that using an EHR system took time away from seeing patients. In addition, EHRs expose private information to hackers (the repeated attacks on hospitals demonstrate how easy is to extort payments for restarting these essential data services); poorly designed interfaces cause endless frustration; and why should every doctor and nurse be a prodigious typist? Above all, what is there to admire about the new model of care with a physician looking at a screen rather than at a patient recounting her problems?

Such lists could be considerably extended, starting with puerile promises of leading alternative lives (as lifelike avatars) in a realistic 3-D virtual space: of course, the most prominent testament to this delusion is Facebook's 2021 conversion by renaming itself Meta and believing that people would prefer to live in an electronic metaverse (I cannot find suitable adjectives to describe this mode of reasoning, if that word is the right noun to describe such an action). Another obvious candidate is the astonishing power of genetic engineering enabled by CRISPR, a new, effective method for editing genes by altering DNA sequences and modifying gene functions: in sensational reporting there is a short distance between this ability and genetically redesigned worlds. After all, has not a Chinese geneticist already begun to design babies, only to be stopped by insufficiently innovative bureaucrats? And just one more recent example: Franklin Templeton's 2022 advertisement that asked "What if growing your own clothes was as simple as printing your own car?" Apparently, the latter (never achieved) option is now considered the template for simplicity. What a perfect solution—when in 2022, even major car makers struggled with getting enough materials and microprocessors for their production lines: just print it all at home!

Instead of extending this list, I will spend a few paragraphs on each of three recently prominent but very different invention categories: drug discovery, long-distance aviation, and AI. Medical research (and associated drug discovery) has become a steady provider of such breakthrough news. As a result of the competitive, grant-supported nature of much modern scientific research, dubious claims begin with the very first announcement of (often preliminary) findings that now take place via press releases by universities or institutions. In 2014 a study of nearly five hundred biomedical and health-related science press releases published in the *British Medical Journal* found that 40 percent of those announcements contained exaggerated advice, a third of them contained exaggerated causal claims, and nearly 60 percent of subsequent news stories based on such releases also contained such exaggerations. Far more remarkably, even completely unsubstantiated claims are now wholesaled as facts and, incredibly, are even approved for use by the very authorities whose duty it is to prevent such a turn of events.

There is no better example of this than the story of Biogen's Alzheimer's drug Aduhelm (aducanumab). In November 2020, eleven members of the US Food and Drug Administration's Peripheral and Central Nervous System Drugs Advisory Committee were asked whether it was reasonable to consider the study submitted by the manufacturer as primary evidence of the effectiveness of the drug for the treatment of Alzheimer's disease. Nobody voted yes, ten members said no, and one was uncertain—and yet seven months later the agency approved the $56,000 per year treatment. Many factors led the panel to its clear negative consensus, including the fact that this treatment approach is based on an established, embedded, but questionable amyloid cascade hypothesis (accumulation and deposition of the beta-amyloid peptide within the frontal cortex and hippocampus in the brain). The hypothesis was formulated in 1984, but all recent clinical trials of anti-amyloid therapies have ended in almost complete failure.

As for the pace of American drug discovery, it has quickened recently, but there has been no rapidly accelerating trend in approval rate. Between 1950 and 1980 the FDA's annual approval rate fluctuated mostly between fifteen and twenty new molecules. It rose above twenty during the 1980s, reached a record of fifty-three in 1996, then fell to seventeen in 2002 before subsequently rising (again, with fluctuations) to a new record of fifty-nine in 2018. The post-2006 rise has been welcome, particularly because the approvals included record numbers of BLAs, or biological license applications: unlike the still dominant NMEs (new molecular entities synthesized by chemists), BLAs are mostly proteins grown and purified from cell cultures of microorganisms (bacteria, yeast) or plant or animal cells, and they have proved effective in treating diseases ranging from rheumatoid arthritis and plaque psoriasis to some cancers.

The first recombinant DNA drug, Humulin, for the management of diabetes, was approved in 1982, and by 2020 more than 170 BLAs had become available, belonging to three categories. Monoclonal antibodies are molecules engineered to restore, boost, or mimic the immune system's attack on alien cells. The first one was approved for dealing with acute transplant rejection, but anticancer and anti-inflammatory therapies are now most common, and in 2020 the FDA authorized two compounds for the treatment of COVID-19. The other two classes of BLAs replace or modulate enzymes (in patients deficient in enzymes able to break down fatty acids or complex sugars) or cell surface receptor functions (used in the treatment of advanced cancers). Until 2013 the absolute number of annually approved BLAs had fluctuated narrowly between two and six, but since then it has been just above ten, a welcome increase but no indication of a sustained and accelerating rise.

Medical research is just one of many research-intensive endeavors that are often presented in too-good-to-be-true press releases. Exaggerated claims of soon-to-come possible practical achievements and not-too-distant commercial deployments have now become the norm in communicating scientific advances to the public. Aviation, whose development I have followed closely for decades, provides a particularly clear example of this trend. In 2017 Boeing and JetBlue funded Zunum Aero, promising nothing less than "transforming US air travel" by 2022 with masses of small (nine- to twelve-person) short-haul electric planes taking off from regional airports. By 2019 Zunum Aero was defunct, but the CEO of Eviation introduced Alice, a nine-seat all-electric commuter plane with a peculiar design of two wingtip pusher motors, at the Paris Air Show in June 2019, claiming that it "is not some future maybe . . . it's operational."

It was not. No test flights were made in 2020, and in 2021 the motors were relocated from the wingtips to the rear of the fuselage. The first flight was promised for late 2021, with commercial deliveries to come in 2024. And just one more aviation claim: on November 8, 2021, Embraer's VP announced that the company (obviously not wishing to be seen as lagging in joining the fashionable quest for zero carbon) was working on four concepts of nine- to fifty-seat aircraft. Yet this was reported as "Embraer Launches a Fleet of 4 New Sustainable Aircraft Designs," while all the company has done so far is to release pictures of four propeller aircraft hybrid-electric accompanied by vague descriptions of electric and hydrogen-electric propulsions that are to enter production in the nebulous 2030s, hardly something that proves it to be "a force to be reckoned with in the race to net-zero." While Embraer is just trying to follow a trend by offering a few conceptual designs, news reporting makes it into "launching a fleet."

But according to two California companies, ZeroAvia and Universal Hydrogen, all of these electric designs will hardly get a chance to prove themselves. ZeroAvia promises to have a superior hydrogen propulsion for a twenty-seat aircraft in service by 2023. Universal Hydrogen not only promises a forty-seat plane powered by green-hydrogen fuel cells (with the fuel delivered in "proprietary, lightweight modular capsules") by September 2022, it even depicts planes for transcontinental and transatlantic range. The latter would have the same passenger capacity as the Airbus 321 but would be about nine meters longer to accommodate hydrogen capsules, which would take up about a third of the fuselage's space in the aft. As simple as that: do a "modest fuselage stretch," fit in a bunch of "lightweight" hydrogen capsules, and be off from JFK to CDG: how come nobody in Airbus or Boeing had such a brilliant idea decades ago? The answer is all too obvious.

But perhaps no category of modern inventions and technical advances has been so poorly and unhelpfully covered as AI. There is no better summary of the challenges faced and failures experienced by AI than *IEEE*

Spectrum's special issue on AI, published in October 2021 (with the main contributions available online). To begin with, the technique's capabilities and goals are often misunderstood even by people engaged in its development. This is not my uninformed opinion but a conclusion reached by Michael Jordan, the world's leading AI researcher at the University of California, Berkeley, who made major contributions to what machines can do—working with human-level competence on low-level pattern recognition skills—but being nowhere near advanced enough to start replacing our brains in reasoning, complex understanding of the real world, and social interactions.

What we have done, often quite effectively, is to deploy some fairly rudimentary analytical techniques to uncover patterns and pathways that are not so readily discernible by our senses but that can be captured, remembered, recalled, and acted upon by computers at scales and speeds unattainable by humans. That is how IBM's Blue beat Kasparov in chess; that is how a program trained on hundreds of thousands of actual x-ray images can discern a cancerous lesion in breast tissue. Unfortunately, as Jordan stresses, "People are getting confused about the meaning of AI in discussions of technology trends—that there is some kind of intelligent thought in computers that is responsible for the progress and which is competing with humans. We don't have that, but people are talking as if we do."

Neural networks, consisting of very large numbers of densely interconnected simple processing nodes (resembling the human brain), are used in machine learning, a process by which a computer learns a task by analyzing training examples. But the progress has been complicated and beset by numerous and sometimes deadly failures. Neural networks are not only brittle (good at specific tasks but deeply deficient in general intelligence, and hence easily overconfident or underconfident in their "judgment") but biased (realities may be far more complex than the training algorithms), prone to catastrophic forgetting, poor in quantifying uncertainty, lacking common sense, and, perhaps most surprising, not so good at solving math problems, even those routinely mastered by high school competitors.

Moreover, training AI systems to achieve a very high level of accuracy, be it in image recognition or in object manipulation, is highly energy-intensive, particularly if one aims at very low error rates. And here are just three additional pithy appraisals by engineers and scientists who have led AI development and who see the achievements and challenges in properly realistic ways. Yoshua Bengio at the Mila-Quebec AI Institute: "I don't think we're anywhere close today to the level of intelligence of a two-year-old child." Yann Lecun at New York University: "What's missing is a principle that would allow our machine to learn how the world works by observation and by interaction with the world." Andrew Ng at Landing AI: "All of AI . . . has a proof-of-concept-to-production gap."

The conclusion is obvious: our quest for AI is an enormously complex, multifaceted process whose progress must be measured across decades and generations and whose impressive achievements on some relatively easy tasks coexist with the much larger realm of intelligence that remains well beyond the capabilities of programmed machines. No matter, in *The Age of AI* the trio of Henry Kissinger, Eric Schmidt, and Daniel Huttenlocher tell us that "the result will be a new epoch" bringing us close to an uncontrollable Armageddon as autonomous weapons will make conflicts both more difficult to predict (as if we've had much success on this score so far!) and to limit. According to this line of thought, living with AI will be an ordeal, while the opposite school of thought sees AI as "immensely helpful," amplifying and optimizing our abilities and ushering in an age of plenty and unprecedented blessings arising from deeply learned neural networks. Unruly complexities and uncertain outcomes find no favor in the modern discourse, which swings between the collapsing civilizations and ever more enticing futures.

At this point I should address the question of progress and innovative speed more directly and support my conclusion regarding the lack of any broad-based rapid exponential growth of inventions with easily verifiable facts. Fortunately, this is not a particularly difficult task. We have plenty of information to contrast the post-1960 advances in computing capacities and speeds with the gains in all other key sectors of modern economies, and the verdict is clear. Rapid exponential growth has been an admirable reality in the advances of solid-state electronics and its applications in devices and designs ranging from personal computers and mobile phones to communication and Earth-observation satellites and data and image processing, but there has been no evidence of any ever-faster innovations in nearly all other sectors of modern economies, from food production to long-distance transportation.

THE MYTH OF EVER-FASTER INNOVATIONS

The pace of innovation, and more generally the rate of any growth, are commonly misunderstood because many people have a mistaken impression of what it means for a variable to grow exponentially. Exponential growth does not mean that every variable whose increase is described by it is growing rapidly. A linearly growing variable increases by the same amount during the same period of time, while an exponentially growing variable increases by the same rate during the same period, and if that rate is very low it will take a long time to see any substantial difference. Here is a real-world example illustrating the difference over time.

During the first two decades of the twenty-first century, Africa's population saw relatively rapid exponential growth averaging about 2.5 percent a year. This means that by 2020 it had grown from about 811 million to 1.34 billion, a gain of 65 percent. Most of the world's undernourished children are in Africa, and milk provides excellent protein for a child's growth—but Africa's average milk yield per cow rose by just 0.8 percent a year during the first two decades of the twenty-first century. Yields of staple cereals rose faster, but their average increase of 1.3 percent was only half the continent's population growth rate. Forget even advances matching the population growth: for a generation, the world's fastest-growing and poorest continent has been falling further behind! Keep this in mind as I contrast rapid exponential growth affecting a minority of phenomena in the real world with the much more common moderate and low rates of exponential growth that govern most human activities and accomplishments.

Nothing has affected, and warped, modern thinking about the pace of invention and the extent of innovation than the rapid exponential advances of solid-state electronics, resulting first in the introduction of transistors (in the late 1940s), then integrated circuits (starting in the early 1960s) and microprocessors (a decade later), followed by similarly rapid increases in their mass-scale deployment in industrial production, transportation, services, homes, and communications. The growing conviction that we have left the age of gradual growth behind began with our ability to crowd ever more components onto a silicon wafer, a process whose regularity was captured by Gordon Moore with his formulation of the now eponymous law that initially ordained a doubling every eighteen months, later adjusted to about two years. As a result, in 2020 we had microchips with seven orders of magnitude (>10,000,000) more components than the first microprocessor, the Intel 4004, released in 1971, did.

These gains provided the foundation for the rapid rise of businesses based on electronic data processing, be they payment schemes (Paypal), e-commerce companies (Alibaba, Amazon), or social media (Facebook, now Meta, Instagram, Twitter). And they have made it possible to go in a lifetime from the 30-centimeter diagonal black-and-white TV screens of the 1950s (embedded in bulky sets) to thin wall-mounted screens with diagonals larger than 200 centimeters able to display millions of colors: going from a 30-centimeter diagonal to a 200-centimeter diagonal gives a viewer a roughly forty-four times larger screen surface. And going from bulky land-line phones (with high long-distance charges) to light portable palm-size mobiles (whose processing power goes far beyond conversations and still images to casual viewing of movies during a subway commute) is a leap whose enormous qualitative difference does not even yield itself to a meaningful qualitative comparison.

And to offer just one classic example of electronics rise, in August 1969, two years before the first microchip appeared, the Apollo 11 computer that guided the capsule to land on the Moon packed just 62 bytes of random access memory (RAM) per kilogram of its (at 32 kilograms clearly nonportable) mass. In 2022 an ordinary Dell laptop used to write this book had about 3.5 gigabytes of RAM per kilogram of its portable (about 2.2 kilograms) mass, or a 1.75 billion-fold gain in performance. Not surprisingly, such stunning gains—so large that most readers would not have noticed had I written million or trillion instead of billion—taking place within such relatively short periods of time leave deep impressions and we notice them far more, and perceive them to be disproportionately more important, than the unchanging or marginally evolving fundamentals of our lives.

Moreover, these admirably rapid exponential gains are seen as harbingers or foundations of similarly impressive gains in other realms of reality. We are told that rapid exponential growth, driven by digitization and advances in AI, already prevails in such fields as solar cells, batteries, electric cars, and even urban farming. And so, in addition to the constant media reporting on waves of stunning inventions, we now have books on exponential technologies and exponential organizations, on general strategies for exponential growth, on the seven essentials to achieve exponential growth (and the eight pillars needed for exponential business growth), and, most sweepingly, on *The Exponential Era: Strategies to Stay Ahead of the Curve in an Era of Chaotic Changes and Disruptive Forces* and on *The Exponential Age: How Accelerating Technology is Transforming Business, Politics and Society*.

At this point we might follow the standard admonition that comes in a soothing voice from the flight deck: just sit back and relax. Everything will take care of itself, unerringly driven by rapid exponential growth that will accelerate, disrupt, transform, and elevate as it ushers in a new era devoid of disease and misery and abounding in material riches. To leave no doubt about what these promises entail, I will quote four proponents of ever-faster growth resulting in an ever more astonishing understanding, endless capabilities, and a surfeit of coming (nearly cost-free) worldly riches: Joel Mokyr, Yuval Harari, Azeem Azhar, and Ray Kurzweil.

Joel Mokyr, an American economic historian, is the most restrained voice in this irrepressibly exuberant quarter. He argues against those who see "the end of invention"—a belief he thinks "is very much alive in our age"—but this belief is not shared by any serious student of either history or science: it is not the end of invention but its recent and future pace that are in dispute, and on these scores, as already noted in the opening chapter, Mokyr belongs firmly to the "ever-faster" contingent. This leads him to forecast the arrival of new antibiotics that would not result in drug resistance among common pathogens, plants "coached" to live with symbiotic bacteria to produce copious fixed nitrogen, the elimination of obesity through the manipulation of "the metabolic factors that determine who will gain weight." Bold yet still rather restrained, he foresees inventive fixes for some long-standing challenges rather than a universal salvation.

In contrast, in his *Homo Deus: A Brief History of Tomorrow,* Yuval Harari portrays a future of unbounded invention in which everything will be known and explained thanks to the mastery of dataism: "*Dataism declares that the universe consists of data flows, and the value of any phenomenon or entity is determined by its contribution to data processing,*" and hence, inevitably, "*We may interpret the entire human species as a single data processing system, with individual humans serving as its chips.*" And if so, then dataism will "provide the scientific holy grail that has eluded us for centuries: a single overarching theory that unifies all the scientific disciplines from musicology through economics to biology." I could not come up with a better retort to this dataistic goulash than David Berlinski's near-perfect verdict: "*Dataism serves chiefly to express Harari's great gullibility. . . . Dataism is not the holy grail . . . it is not to unify anything. . . . Men are not about to become like gods. Harari has been misinformed.*" Indeed, and grossly so!

Azeem Azhar—an entrepreneur, investor, creator of the newsletter *Exponential View,* and author of *Exponential: How Accelerating Technology Is Leaving Us Behind and What to Do about It,* is even more infatuated with the ascent of the machine as he sees new technologies "*being invented and scaled at an ever-faster pace, all while decreasing rapidly in price.*" He includes in that group not only computing, AI, and biotech but also renewable electricity and energy storage. As a result, a cornucopian world is just around the corner: "*We are entering an age of abundance. The first period in human history in which energy, food, computation and much else will be trivially cheap to produce.*" This reminds me of what I heard in grade school under the Evil Empire when our rulers were promising a similar kind of earthly nirvana as soon as they were done with building communism.

Reasoning with true believers—be they of religious or ideological persuasion or cornucopian techno-optimists—is not an option, but there is one thing I could muster in defense of Harari and Azhar. They—unlike Ray Kurzweil, the most assiduous proponent of exponentially accelerating innovation—have not put any firm dates on the arrival of this all-knowing, all-explaining, abundance-delivering (essentially at no cost!) earthly state of affairs, and they have also kept the actual innovative speed unspecified. Kurzweil, an American inventors and futurist and now Google's director of engineering, has no doubts on either account. According to him, 2045 is the year when machine intelligence will have surpassed human intelligence, when these two entities will merge and we will become immortal, making the colonization of the universe a rather simple task as—the inevitable consequence of ever-faster exponential growth, ending in the Singularity—knowledge, expanding in all directions, will be filling the universe at the speed of light.

Except that it will not. The rapid exponential growth emblematic of many microprocessor-enabled activities and companies that offer such services to the public has already entered a more moderate expansion stage. Printing with ever-shorter wavelengths of light made it possible to crowd in larger numbers of thinner transistors on a microchip: the process began with transistors 80 micrometers wide; 7-nanometer–based chips are now common (their width is only 0.0000875 that of the first design), and in 2021 IBM announced the world's first 2-nanometer chip, to be produced as early as 2024. Because the size of a silicon atom is about 0.2 nanometers, a 2-nanometer connection would be just ten atoms wide, and the physical limit of this fifty-year-old reduction process is obviously in sight.

Between 1993 (Pentium) and 2013 (the AMD 608), the highest single-processor transistor count went from 3.1 million to 105.9 million, the final total being actually a bit higher than prescribed by Moore's law (doubling every two years would bring it to 99.2 million). But progress has slowed. In 2008 the Xeon had 1.9 billion transistors and a decade later the GC2 packed in 23.6 billion, whereas a doubling every two years should have brought the total to about 60 billion. As a result, the growth of the best processor performance has slowed from 52 percent a year between 1986 and 2003 to 23 percent a year between 2003 and 2011 and eventually to less than 4 percent between 2015 and 2018. As with all cases of growth, an S-curve has been forming, and the period of very rapid exponential growth is history.

More important, that much-admired post-1970 ascent of electronic architecture and performance has no counterpart in nearly all other aspects of our lives: rapid exponential growth has not marked the advances in either fundamental economic activities on which modern civilization depends for its survival—ranging from crop yields to efficiency gains in energy uses, from transportation speeds to the ability to design and complete large engineering projects—or the critical determinants of health and quality of life, including the rate of new drug discoveries and gains in longevity. Examples of these realities abound.

The Instagram app attracted 25,000 users the day it was launched and had ten million of them within ten weeks: that is, obviously, the result of a dizzying but inevitably only temporary exponential growth: unless it begins to communicate with numerous extraterrestrial civilizations, Instagram cannot sign up more users than Earth's population. And is the fact that Instagram was sold to Facebook for more than $1 billion when it still had just thirteen employees a swoon worthy example of exponential growth or a perfect example of the irrational priorities of modern society? Just check the valuations of companies producing milk or bread or tomatoes: while you cannot live without a ceaseless supply of food, hundreds of millions of people would not notice the instant demise of Instagram or TikTok.

And what is more important, Instagram's temporarily dizzying rise or the fact that even as it was taking place, the global share of people who remain undernourished was increasing? That share was declining, slowly and nearly linearly, for a generation, to a low of 8.3 percent of the world population by 2015, but it has since risen again to about 10 percent. Moreover, after rising 4 percent in just three years, the share is now about 20 percent in Africa: every fourth person experiences hunger in the sub-Saharan part of the continent, and nearly every third in its center. Yet during the coming generation more than 90 percent of the world's population growth will take place in already hungry Africa, and we know that undernutrition among pregnant women and growing children is, in so many ways, a lifelong sentence, that it deprives adults of their full working capacity, and that it reduces everybody's quality of life.

No matter whether we look at the increases in staple grain yields required for the survival of now eight billion (soon to be nearly ten billion) people or at the performance of processes indispensable for the functioning of modern civilization, we see no signs of any rapid exponential advancement. Moore's law, with the doubling of the microprocessor performance approximately every two years (and faster in the earliest period), implies a high rate of annual exponential increase (about 35 percent, even faster in the earliest period of development), and in fifty years of improvements that has resulted in a gain of seven orders of magnitude (that is, more than 10,000,000,000 times greater). In contrast, annual gains in our food, materials, and energy production have been only a small fraction of that, resulting from very low rates of exponential growth, mostly on the order of 1–2 percent a year; the first rate increases the initial value only 1.65 times in fifty years, while exponential growth of 2 percent a year will have an outcome 2.7 times higher after half a century.

Here are some notable recent crop yield outcomes. During the first two decades of the twenty-first century Asian rice harvests increased by 1 percent a year, yields of sorghum, sub-Saharan Africa's staple, went up by only about 0.8 percent a year, and in 2020 the average yields of both Australian wheat and European potatoes were a mere 1 percent higher than two decades earlier, implying a minuscule (less than 0.1 percent) annual growth rate. And, unfortunately, many similarly low growth rates prevail in animal production: I have already contrasted Africa's relatively rapid population growth with the continent's hardly discernible gains in milk production per animal.

Similarly low exponential growth rates characterize the economic growth of many countries that have the greatest need to advance. Since 1960 the average per capita gross domestic product of sub-Saharan Africa has been growing annually by no more than 0.7 percent when expressed in constant monies. In Brazil it has been less than 2 percent for half that time, while in exceptionally fast-growing China it was above 5 percent between 1991 and 2019. Growth rates of technical advances, productive capacities, and efficiencies have been similarly restrained. Most of the world's electricity is generated by large steam turbines whose efficiency got better by about 1.5 percent per year during the past hundred years. We keep making steel more efficiently, but the annual decline in energy use in the metal's production averaged less than 2 percent during the past seventy years. And, as already noted (and setting aside the failed Concorde), the average speed of jet flight has not seen any increase since 1958.

During the past decade, observant readers have seen many news items about stunning breakthroughs in battery designs, but I cannot find any ever-accelerating growth in the performance of these portable energy storage devices in the past fifty years. In 1900 the best battery (lead-acid) had an energy density of 25 watt-hours per kilogram; in 2022 the best lithium-ion batteries deployed on a large commercial scale (not the best experimental devices) had an energy density twelve times higher—and this gain corresponds to exponential growth of just 2 percent a year. That is very much in line with the growth of performances of many other industrial techniques and devices—and an order of magnitude below Moore's law expectations. Moreover, even batteries with ten times the 2022 (commercial) energy density (that is, approaching 3,000 Wh/kg) would store only about a quarter of the energy contained in a kilogram of kerosene, making it clear that jetliners energized by batteries are not on any practical horizon.

Much has also been written about a reverse manifestation of exponential change, about the impressively declining cost of solar photovoltaic cells leading to near-miraculous breakthroughs in solar electricity generation. The latter claim has been particularly popular: I encourage you to check those breathless reports of constantly and rapidly falling photovoltaic (PV) cell prices, and you will see how, if they were the only determinant of the actual cost of PV generation, we would soon be arriving at almost the same place where nuclear generation claims began in the mid-1950s, with solar generation being too cheap to meter, indeed, being absolutely a free give-away.

In reality, detailed US data for residential PV systems (twenty-two panels) show that the module cost is now only about 15 percent of the total investment. The rest is needed to cover structural and electrical components (panels must be mounted on supports on roofs or on prepared ground), inverters (to change the direct current to alternating current), labor costs, and other soft costs. Obviously, none of these components, from steel and aluminum to transmission lines, permitting, inspection, and sales taxes, is tending to zero, and hence the overall costs of installation (dollars per watt of direct current delivered by the panels) show a distinctly declining rate of improvement: between 2010 and 2015 they fell by 55 percent, between 2015 and 2020 by 20 percent. And these costs do not include the additional outlays that will have to be made with the increasing share of intermittent sources (solar and wind) in overall electricity generation.

To prevent extended shortages and supply interruptions, either these modes of electricity generation will have to be backed up by sufficient on-demand reserves or the regions dependent on solar and wind supply will have to have reliable long-distance high-voltage transmission links to bring electricity supplies from places not affected by temporary heavy cloudiness or extended calm periods. The costs of this entire electricity-supply system have not been declining, and the construction of long-distance high-voltage transmission lines necessary to provide large-grid security has been falling behind the planned needs, both in the US and in Europe. The real cost of PV panels should also include their dismantling and disposal or, preferably, their recycling. And if the costs of renewable electricity generation have been plummeting, why do the three EU countries—Denmark, Ireland, and Germany—with the highest share of energy from new renewable sources, wind and solar, have the continent's highest electricity prices? In 2021 the EU mean was €0.24/kWh, but the Irish price was 25 percent higher, the Danish price 45 percent higher, and the German price 37 percent higher.

No matter. Such questions, reminders, and objections—referring to basic physical realities, known constants, available rates, and capacities—are now seen as almost irrelevant, nothing but challenges to be vanquished by ever-accelerating innovation. But there are no signs of such a sweeping acceleration; there is no indication of ever-faster inventions as far as the most fundamental human activities are concerned. This inevitable conclusion is now supported by a detailed study of innovation across American industries spanning nearly two centuries, from 1840 to 2010. Its authors, four American economists led by Bryan Kelly, used textual analysis of patent documents to create new indicators of innovation and to identify breakthrough innovations as the most significant patents in order to construct indices of long-term change across all major industries.

This analysis captured the evolution of innovation waves and provided unambiguous quantitative support to the previously reached conclusions regarding the timing of the most fundamental innovations that have created the modern world. Breakthrough patents in the furniture, textiles, and apparel industries, in transportation equipment, machinery manufacturing, metal manufacturing, wood, paper, and printing, and in construction all peaked before 1900. Mining and extraction, the coal and petroleum industries, mineral processing, electrical equipment production, and plastics and rubber products had their innovative waves and peaks before 1950, and the only industrial sectors with post-1970 peaks have been agriculture and food (the wave dominated by genetically modified organisms), medical equipment (from MRI and CT scanners to robotic surgical tools), and, of course, computers and electronic products.

These incontrovertible findings refute any assertions about an ever-increasing rate of innovation and put claims about the extraordinary impact of recent inventions into a proper historical perspective. Perhaps the best way to appreciate this reality is to try to imagine the world without the benefits brought by the latest wave of innovation, which brought fundamental breakthroughs in computers and electronics: a world without microprocessors, without ubiquitous computing, and without any social media. To do so is quite easy as that was the world of the early 1970s: Intel's first microchip was designed in 1971, but its first 16-bit microprocessor, the 8086, was released only in 1978; Microsoft was established in 1975, but the first mass-produced personal computer, the IBM PC, came in 1981.

In the absence of these solid-state components and devices, the world of the early 1970s was one of new high-yielding wheat and rice cultivars, of efficient gas turbines (stationary in electricity generation, and powering wide-body jetliners), of large container ships, of growing megacities, of telecommunication and weather satellites, and of antibiotics and vaccines. All too obviously, a high-energy, high-quality-of-life affluent civilization is not based on post-1971 electronics: the development and diffusion of electronics have been welcome and helpful and valuable, but most definitely not fundamental.

Then reverse the task, and try to imagine today's electronics-based world running without large-scale electricity generation, without high-yielding agriculture, without the dominant prime movers (engines, turbines, electric motors), and without the mass production of materials ranging from inexpensive steel, nitrogen fertilizers, and aluminum to even lighter plastics. None of these fundamental components of modern civilization is predicated on the widespread reliance on solid-state electronics, indeed, not on its very existence: its diffusion has made most of these processes easier to manage, monitor, and improve, but they existed for decades before the arrival of the late twentieth-century solid-state-based electronics.

And the historical corrective goes even further, as the energetic and material foundations of modern civilization go back into the five decades before the beginning of World War I and, to a surprisingly high degree, to a single decade, the 1880s. That decade saw the invention and patenting, and in many cases also the successful commercial introduction, of so many processes, converters, and materials indispensable for modern civilization that their aggregate makes the decade's record unprecedented, and most likely unrepeatable. Bicycles, cash registers, vending machines, punch cards, adding machines, ballpoint pens, revolving doors, and antiperspirants (and Coca Cola and the *Wall Street Journal*) could be dismissed as the decade's minor inventions and innovations.

Above all, the inventions of fundamental and lasting importance included the near-complete creation of the system of electricity generation, distribution, and conversion. The decade saw the world's first coal-fired and hydroelectric plants, steam turbines (the mainstay of thermal electricity generation), transformers, transmission (both of direct and alternating current), and meters, and electricity was used by the newly invented incandescent light bulbs, electric motors, and elevators, as well as for welding, urban transportation (street cars), and the first kitchen gadgets. Our microchip-rich world depends on a reliable electricity supply, and by 2020 thermal and hydroelectric generation still provided more than 70 percent of all electricity, with the new renewable sources, wind and solar, contributing only about a tenth of that.

The 1880s were also the decade when three German engineers invented motor cars powered by internal combustion engines, when a Scotch inventor came up with inflatable rubber tires, an American chemist with the way to produce aluminum, and an American architect to complete the world's first multistory steel-skeleton skyscraper. The enduring and fundamental importance of these inventions is self-evident. And there was still more: between 1886 and 1888 Heinrich Hertz proved that James Clerk Maxwell was right, as he generated and transmitted electromagnetic waves, measured their frequencies, and correctly placed them between "the acoustic oscillations of ponderable bodies and the light-oscillations of the ether." This is where the modern world of intangible wireless communication began, with mobile phones and social media being what I have called the fifth-order derivations of Maxwell's ideas (Hertz being the second, the earliest pre–World War I broadcasts the third, the mass diffusion of vacuum tube–based electronics the fourth, and solid-state electronics the fifth).

WHAT WE NEED MOST

The historical verdict is indisputable: without inventions and the ensuing innovation, modern societies could not have achieved their high quality of life, including unprecedented longevity, affluence, education, and high mobility. The cumulative, combined effects of inventions reached new highs after the middle of the nineteenth century (both as far as their quantity and their transformative qualities were concerned) and they were further enhanced during the twentieth century, the time of extraordinarily wide-ranging innovation that has extended the benefits of the most consequential inventions (ranging from antibiotics and synthetic fertilizers to inexpensive steel and affordable electricity) to most of the world's population, now approaching eight billion people.

Obviously, we will need many new inventions to tackle many persistent unresolved problems and deal with new challenges. As with any list, you can turn for guidance to the internet, but the choices will mostly be just pathetic click-bait, spinning you between trivia and rank science fiction. One list in the latter category, looking at concepts that have yet to become realities, includes the "edible Jell-O squishable cup" and "levitating cloud-shape sofa"—but even "serious" lists are full of completely unnecessary frivolities or sci-fi-type wishes: do we really need to be directly reading other people's minds, communicating with extraterrestrials, or live forever? As for the appeal of the last option, readers unfamiliar with Jonathan Swift's writings should consult his description of the immortal Struldbruggs in Luggnagg (*Gulliver's Travels*) to consider the dubious benefits of that achievement.

But could not we come up with a manageable number—say, a score or two—of the most desirable items based on the two overriding needs: to improve the fundamentals required for dignified life of the world's population, and to do so without excessive impacts on the biosphere? In physical terms, this means securing adequate supplies of food, water, energy, and materials needed to lead healthy lives with decent life expectancies;

in mental, social, and economic terms it would mean ensuring the opportunities for education and employment and providing generally accessible, good-quality health care; and all of that should be done while leaving sufficient resources for the long-term survival of other species—even as the total number of the human species is still increasing.

While that might be a reasonable framework for defining the selection of the most desirable inventions, it is obvious that because there could be no universal metric to assess their impact once they became reality, there could be no clear ranking of the need for such breakthroughs, not even their grouping into relatively similar categories. We could measure health, longevity, and quality-of-life gains by using the common denominators of life-years saved (LYs) or quality-adjusted life-years (QALYs) gained. The QALY concept was developed to combine length of life and quality of life into a single index number that could be used for comparisons of outcomes as well as a common denominator for costs. But how do we compare a much-sought breakthrough in treating certain intractable cancers with breakthroughs in crop genetics, electricity storage, or steel production?

Obviously, all of these contribute to quality of life: QALY gains would be impossible without better nutrition, a reliable electricity supply, and numerous irreplaceable uses for steel products—but there is no common metric by which to judge their relative importance or rank them as to their indispensability: the complexity of modern societies, with its now overwhelming densities of links and feedbacks preclude that. And simple, unranked lists of ten or thirty most wanted items might not be any better: if done by individuals they would betray inevitable personal predilections and biases, and groups charged with the task might find it impossible to come up with a clear consensus within the given limits. Consequently, perhaps the best thing I can do is to explain the magnitude of inventive tasks we face while reiterating the key lesson of this book: the exponential growth of microprocessor capabilities and the devices defined by them, ranging from computers to mobile phones, is an exception, not the norm dominating the recent waves of inventions.

I will illustrate these challenges by using two very different examples, by looking back at half a century of a focused, well-supported inventive quest, to reduce cancer's toll in modern society, and by looking ahead at the prospects of the slowly unfolding process of decarbonization, the transition from fossil fuels to energies whose production and conversions do not emit carbon dioxide and methane, the two leading greenhouse gases implicated in anthropogenic global warming. By no means I am trying to imply that the pace of future reduction of global CO_2 emissions will resemble that of lowering cancer mortality. I am merely using the well-documented history of one inherently complex endeavor to indicate the likely challenges of another complex (though qualitatively and quantitatively very different) transformation that will not be possible without major new inventions.

As already noted, on December 23, 1971, President Richard Nixon signed the National Cancer Act, launching the series of government-sponsored programs that became known as the war on cancer. This was an unfortunate metaphor, as if a time-limited assault could succeed in vanquishing more than a hundred types of the disease, including many gender- and age-specific forms. The original mandate was simply to "support research and the application of the results of research, to reduce the incidence, morbidity and mortality from cancer." The act set no time-frame for achieving specific goals, but the goal of eventual eradication was implied when Nixon compared the quest to the successful Moon landing that took place just two years before he signed the act. More than three decades later, in 2003, Andrew von Eschenbach, at that time the director of the National Cancer Institute, called to "eliminate the suffering and death from cancer, and to do so by 2015"—and President Obama spoke about finding "a cure for cancer in our time."

Scientists and physicians best informed about the challenges of this endeavor have always appreciated that this was not primarily a matter of financing the development of new drugs or devising new treatment procedures. What was required in the first place was substantial advances in the basic scientific understanding of carcinogenesis, heritability, and disease progression. Inevitably, uncovering these fundamentals is a prolonged process, and, not surprisingly, the retrospectives looking at the first twenty-five years of the "war on cancer" were dominated by cautious optimism rather than by any recital of triumphs. By 1996 impressive advances were being made in treating and curing leukemias (acute lymphocytic leukemia in children saw the most gratifying retreat) and lymphomas, but it was clear that the NCI's target of halving the cancer mortality by the year 2000 could not be achieved. In fact, the overall cancer mortality kept rising until 1991, when it reached 215 per 100,000 population, and the prognoses for patients diagnosed with advanced metastatic cancers were only marginally better than in the early 1970s.

After falling since 1991, the overall cancer mortality in 1999 was the same as in 1975, but then, finally, came a period of steady reductions. Between 1999 and 2019 the American age-adjusted cancer death rate dropped by 27 percent, from about 201 to about 156 deaths per 100,000 people, with the drop more pronounced among males (31 percent) than among females (25 percent) but with cancers remaining still more common among men (173 per 100,000) than among women (126 per 100,000). Age adjustment is an imperative part of any historical comparisons because cancer death rates rise with age (from about 10 per 100,000 for people in their early thirties to just over 200 per 100,000 for people in their late fifties) and because the populations of affluent countries have been steadily aging.

The most important new basic science advances and treatments that contributed to declining mortality began with the discovery of the first oncogenes (cancer-causing genes), the most commonly mutated gene in human cancer, and the approval of tamoxifen, an antiestrogen drug to treat breast cancer, during the 1970s. A new oncogene associated with the more aggressive forms of breast cancer as well as the link between human papillomavirus and cervical cancer were discovered in 1984. A decade later came the cloning of tumor suppressor genes to fight breast and ovarian cancer, and during the late 1990s the FDA approved the first monoclonal antibodies to treat non-Hodgkin lymphoma (rituximab) and metastatic breast cancer (trastuzumab). The first vaccines against human papillomavirus were introduced in 2006 and 2009, and in 2010 came the first human cancer treatment vaccine using a patient's own immune system to limit metastatic cancer.

After 2010 came new monoclonal antibodies to treat advanced melanoma, breast cancer, and various solid tumors, and the first personalized treatment (removing a patient's specific cells, genetically altering them, and then infusing them back to stimulate the immune system to attack cancer cells) for one type of leukemia. These advances in treatment were accompanied by more widespread screening and early diagnoses, and they helped produce some substantial increases in five-year survival rates compared to the mid-1970s: most impressively, from 47 to 74 percent for non-Hodgkin lymphoma, from 75 to 91 percent for breast cancer, and from 82 to 94 percent for melanoma. But major site differences remain: pancreatic cancer's five-year survival rate tripled, but it is still only 9 percent; the esophageal cancer survival rate has more than quadrupled, to 21 percent; while 98 percent of patients with thyroid cancer have survived longer than five years. And despite the declining rate of smoking, lung cancer remains the leading malignancy (even among females it is about 45 percent more common than breast cancer), and its five-year survival rate rose from 12 percent to only 20 percent.

And the "war" continues, now under a different label. In February 2022 President Biden reignited "Cancer Moonshot to End Cancer as We Know It," although this headline on the White House website was followed by a more realistic qualifier: "Biden-Harris Administration Sets Goal of Reducing Cancer Death Rate by at least 50 Percent Over the Next 25 Years, and Improving the Experience of Living with and Surviving Cancer." At the same time, the rising US mortality caused by drug overdose is a perfect reminder of the fact that the hard-won gains could be largely negated by mounting losses elsewhere. American drug overdose deaths totaled about 48,000 in 2015, but in the twelve months ending in April 2021 they had doubled, to about 98,000, compared to about 320,000 deaths from all cancers and 142,000 deaths from lung cancer. Given the age difference of the two kinds of mortality—overdoses occur mostly among people less than forty years old, while cancer deaths occur mostly among people older than fifty—the recent rise in drug-related deaths might have completely negated the years of life gained with the latest cancer treatments.

Even this very brief review makes it clear that the earlier calls for relatively rapid eradication of cancer were quite unrealistic and that the quest for a substantial reduction in cancer mortality is a prolonged, multidecadal, intergenerational process with uneven outcomes for cancers in specific body sites. Another testimony to this reality is that in 2016 the US Congress passed the 21st Century Cures Act whose goal is to accelerate treatment and deliver new innovations to patients faster and more efficiently. And I find many of the experiences gained from the war on cancer to be highly applicable to other endeavors whose nature and goals may be quite different but whose inherent complexities and overall scale of eventual accomplishment are comparably daunting.

These are the foremost generic lessons: basic (scientific and technical) understanding must precede specific applications (perhaps the most obvious but repeatedly ignored reality); critical variables may get worse before they get better; it is unwise to specify outcomes by dates; even near-term targets, no more than ten years away, will be missed; some very impressive advances will take place alongside barely changing realities; intra- and international differences (for a variety of reasons) will continue to be significant; initial cost estimates will escalate; and the gains may be partially negated by new developments, undermining the hard-won achievements.

All of these lessons are perfectly applicable to any realistic assessment of our chances to accomplish relatively rapid global decarbonization. To begin with, our needs for much-expanded basic scientific understanding and for the ensuing waves of new inventions needed to drive global decarbonization are much greater than has been generally acknowledged. This quest—on a global scale and involving masses measured in billions of tons—is at the opposite end of the size spectrum represented by molecular cancer therapies, but it too will need steady waves of inventions. As Bill Gates noted in October 2021, "Half the technology needed to get zero emissions either doesn't exist yet or is too expensive for much of the world to afford." Obviously, to remedy these gaps will require unprecedented efforts in inventing new modes of energy extraction, storage, and conversion ranging from the production of green hydrogen (this gas is now made solely by reforming fossil fuels, natural gas, and, to a much lesser extent, coal) to mass-scale high-energy-density storage of electricity.

The latter need is particularly urgent because the unfolding energy transition to carbon-free electricity (dominated by wind, solar photovoltaics, and solar central power) and carbon-free fuels (hydrogen, ammonia, synthetic fuels made from captured CO_2) would greatly benefit from new, superior ways of large-scale electricity storage. But even if we got batteries whose energy density was an order of magnitude higher than today's best lithium-ion batteries, their energy density would still be less than a quarter of the energy density of the refined liquid fuels (gasoline, kerosene, diesel fuels) that now dominate all forms of transportation. Moreover, new high-energy-density batteries would also need to attain unprecedented capacities in order to store enough electricity to supply megacities at times when wind and solar generation will not be available (Asian megacities repeatedly visited by typhoons are the best example of these enormous storage needs).

That global warming will get worse before it gets better is a foregone conclusion: even an instant (and totally theoretical) cessation of all greenhouse gas emissions could not bring an instant stabilization and decline of the average tropospheric temperature. The predilection of grand global meetings (as well as of national strategies) to set decarbonization targets at years ending in a zero or five (45 percent less carbon by 2030 globally; no carbon emissions from US electricity generation by 2035; net zero carbon globally by 2050) is an obviously arbitrary exercise and meeting these goals would require extraordinary technical and economic transformation on the global scale.

A few examples illustrate the wishful nature of such targets. In the year 2000 fossil fuels supplied 87 percent of the world's primary energy, while in 2020 that share was 83 percent, hence an annual reduction of 0.2 percent—but now we are told that we should end our dependence on carbon by 2050. But going from 83 percent to zero in thirty years would require cutting 2.75 percent of global fossil carbon every year, a rate nearly fourteen times faster than we managed during the first two decades of the twenty-first century. Where are the technical capabilities and financing that would allow as to realize, instantly, such a large annual cut and sustain it for three decades?

Just a couple of examples arising from the targets announced at the November 2021 UN Climate Change Conference (COP26) makes it clear how extraordinarily unlikely it is that they could be realized. The latest goal is to reduce global CO_2 emissions from the combustion of fossil fuels by 50 percent by 2030 relative to the 2010 level of 30.4 billion tons. This means that during the nine years between 2022 and 2030 we would have to reduce them by about 13.7 billion tons or by an annual linear decline averaging about 1.5 billion tons (fig. 5.1). Let us assume that

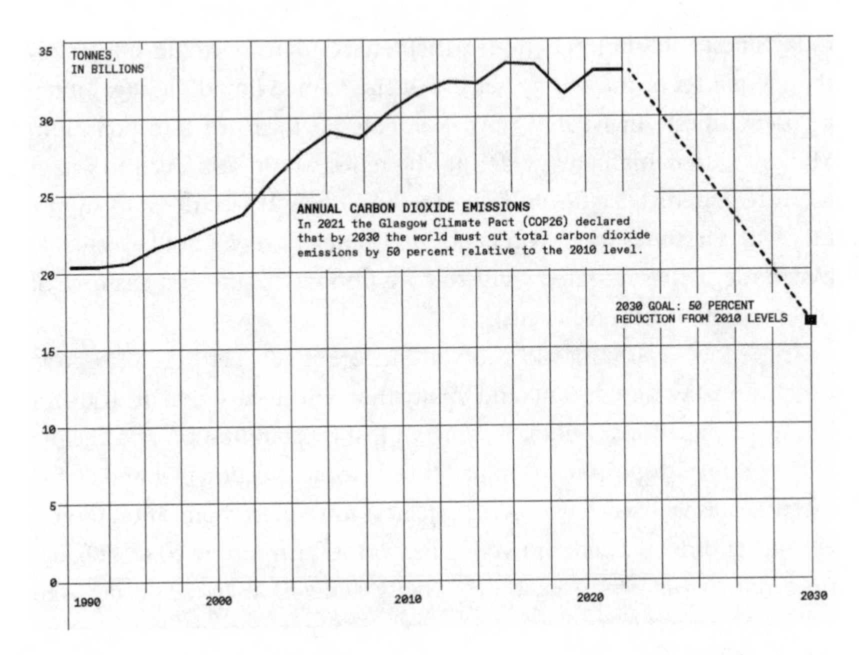

Figure 5.1 Global decarbonization by 2030. *Source:* Vaclav Smil, "Decarbonization Algebra," *Spectrum* February 2022; data from International Energy Agency and UN Framework Convention on Climate Change.

all energy-consuming sectors will share these cuts equally and that the global energy demand will not grow (in reality, during the pre-pandemic decade it was increasing by 2 percent a year).

In 2019 the world produced 1.28 billion tons of pig (cast) iron in blast furnaces fueled with coke made from metallurgical coal. That pig iron was charged into basic oxygen furnaces to make about 72 percent of the world's steel (the rest comes mostly from electric arc furnaces melting scrap metal). As of 2022 there is not a single commercial steelmaking plant reducing iron ores by hydrogen. Moreover, nearly all hydrogen is now produced by the reforming of natural gas, and zero carbon iron would require as yet nonexistent mass-scale electrolysis of water powered by renewable energies. A 40 percent cut in today's carbon dependence would mean that by 2030 we would have to smelt more than half a billion tons of iron—that is more than today's annual output of all of the world's blast furnaces outside China—by using green hydrogen instead of coke. What are the chances of that?

In 2021 there were some 1.4 billion motor vehicles (about 1.2 billion cars, SUVs, pickups, and vans and 200 million buses and trucks) on the road, of which fewer than 17 million (only about 1.2 percent) were electric and 99 percent were powered by gasoline or diesel fuel. Even if the global road fleet did not grow, having 40 percent of it decarbonized by 2030 would require about 570 million new electric (or hydrogen- or ammonia-fueled) vehicles made in nine years; that is about 63 million a year, or more than the total global production of all cars in 2019, and all electricity to produce those fuels would have to come from zero-carbon sources. What are the chances of that?

Inevitably, these targets will not be met (an unprecedented collapse of the global economy might be the only way to do that), and while the progress toward the complete decarbonization goal will be (as it is already) much faster than the global mean in some smaller countries with an abundance of opportunities for renewable conversions (Norway, Iceland, Denmark, Finland), many large, populous, and still low-income economies (India, Pakistan, Indonesia, Nigeria) will move much slower. As for the costs, we are now only at the very beginning of a long transition (new noncarbon sources of energy supplied less than 7 percent of the total demand in 2020), and while we can count on some specific conversions to become, as some already have, substantially cheaper, at this stage nobody can offer good cost estimates of developing entirely new infrastructures (such as the green hydrogen generation, transportation, and storage to replace billions of tons of crude oil and natural gas) on a global scale.

And the quest for decarbonization also offers perfect examples of gains being partially negated not only by other concurrent developments but also by the very process of expanding noncarbon conversions, with wind-powered electricity generation being the most obvious case. The construction of massive wind turbines requires considerable quantities of reinforced concrete (cement and steel) for the foundations, steel for towers and nacelles, plastics for large blades, and lubricating oils for smooth motor service; the turbine parts are transported to their onshore or offshore sites by large trucks, ships, and tugs, and offshore sites are often serviced by helicopters. All of these components and delivery means rely heavily on fossil carbon, either as fuel (to make steel, cement, and plastics, to power vehicles and ships), feedstock (to synthesize plastics), or lubricants—and if, for example, the wind electricity were to displace a large share of today's coal-fired generation, the need for these fossil carbon inputs would rapidly multiply.

This dependence would be eliminated only if all those productive and transportation processes (from steel- and cement-making to trucking and lubrication) were supplied by noncarbon energies, including the smelting of iron without coke (by relying on hydrogen), deriving feedstocks from biomass (rather than from hydrocarbons), and using only electric or hydrogen-fueled transportation and synthetic lubricants. There is no need to have a deep engineering grasp of these realities to know that such a completely carbon-free outcome will require many decades of gradual progress. Moreover, this reality means that the faster we move to adopt noncarbon energy-producing processes, the more we will have to rely on carbon-based production and transportation methods that cannot be replaced rapidly with noncarbon processes even if those were readily available—and in most cases they are not.

Global aviation offers yet another perfect example of these nonexistent alternatives. According to the Glasgow Climate Pact, by 2030 the world should cut its CO_2 emissions by 45 percent relative to the 2010 level. This translates to about a 40 percent cut in global emissions relative to 2021 total (after the 2020 pandemic-induced drop they had nearly recovered to the 2019 level). But how could we cut the emissions in commercial aviation, now totally dependent on kerosene, by two-fifths in just nine years? Our best commercially available batteries have an energy density on the order of 300 Wh/kg, whereas aviation kerosene contains more than 12,000 Wh/kg.

This is more than a forty-fold difference, and it would require a miracle to have commercially available batteries with just a half or a third of kerosene's energy density before 2030. Similarly, there is not a single commercial hydrogen-powered airplane in service anywhere, and the well-known challenges of storing this (in its liquid form cooled down to –253°C) high-energy-density fuel aloft make it extremely unlikely that we will see fleets of hydrogen-powered airliners even by 2040—but a 40 percent carbon cut by 2030 would require us to have some 10,000 non-kerosene (electric or hydrogen) aircraft in service by 2030 (the global fleet is now about 25,000 aircraft) in order to fly about 1.8 billion zero-carbon passengers a year. Clearly, even an unprecedented explosion of inventions is not going to make that happen,

But there is yet another way how to look at which inventions are needed most, with the priorities dictated by changing the prevailing state of affairs. This means striving for significant reductions of existing inequalities or for at least narrowing the health, education, and income gaps, above all the most conspicuous differences between one billion people in affluent economies and the more than three billion people surviving essentially at the subsistence level, with recurrent morbidity, premature mortality, and hence shortened life expectations. Meeting the essential water, food, energy, and material needs would then come first.

We need cheaper, less space-intensive (also modular), and more effective water treatment techniques leading all the way to near-complete recycling, and we also need more desalination. In field agriculture we need higher yields in countries where most of the nearly one billion of currently undernourished people now live, and reducing this total will also need more equitable access to available food and to the supply of micronutrients whose shortages affect many disadvantaged populations and can be remedied at very low cost. As in the case of undernutrition, nearly a billion people still have no access to electricity, and the average annual per capita energy use among more than three billion people (40 percent of the global population) is less than 25 gigajoules per capita, that is, at levels comparable to those of well-off European countries and North America during the middle of the nineteenth century! Obviously, we need to raise those dismally low access rates and consumption averages.

Undoubtedly, all of these desiderata would benefit from new inventions, but effective and relatively rapid progress in the right directions does not depend on them. Satisfying the water and food requirements of the entire global population does not depend on any new spectacular inventions (as all key components are already considerably advanced, and in some places have been reliably operating for decades) but rather determined innovation that would diffuse these benefits and reduce their costs. The same is true about electrification and about raising the average levels of primary energy use. And this list of desirable basics can be easily extended by items ranging from antibiotic resistance to improvements in education.

Again, many inventions might be useful, but we know what we should have done, and should be doing. To limit the diffusion of antibiotic-resistant bacteria we must prescribe antibiotics with care (and not overuse them, as is the norm in affluent countries, or sell them without prescription, as is common in many low-income countries, or now universally on the internet) and not allow preventive mass dosing of domestic animals, which now receive an order of magnitude more antibiotics than humans do. As for the good foundations of universal education, we know that that can be achieved without every child having a computer or without extraordinarily high spending. Just compare the results of international math and science testing, with the US ranking below Poland on all three scores, reading, math, and science, despite spending 2.5 times more per student on grade school education and three times as much on high school education.

We know perfectly well how to remedy all of these undesirable or outright demeaning realities without any brilliant inventions but with the determined extension of known and reliable methods, skills, and procedures. In the grand scheme of things, improving what we know and making it universally available might bring more benefits to more people in a shorter period of time than focusing overly on invention and hoping that it will bring miraculous breakthroughs. To forestall the obvious critique, this is not an argument against the determined pursuit of new inventions, merely a plea for a better balance between the quest for (perhaps, but not assuredly) stunning future gains and the deployment of the well-mastered but still far from universally applied understanding and achievements.

Perhaps it all comes down to personal preferences, and I have always felt strongly about doing first things first. And that means, to choose two notable examples, doing away with the micronutrient deficiency blighting the lives of hundreds of millions of children before deploying supersonic transport. At the same time, I have always been a realist and a skeptic, and I know that resources for invention and innovation are never allocated on the basis of such rationally comparative needs, and that my priority plea could be assailed as misplaced and insufficiently ambitious or aspirational. Moreover, it might be easier, for many reasons, to support quests for even dubious inventions rather than to carry on with alleviating human misery.

In any case, we are not going to stop inventing new materials, products, processes, and procedures and this means that we will have to keep reckoning not only with inevitable design failures stemming from unprecedented challenges and from the lack of experience but also with repeated, and major, failures resulting from human preferences, priorities, biases, and irrational attachments to certain quests. In that sense, and contrary to mistaken claims of the ever-faster pace of invention, *nihil novi sub sole*.

FURTHER READING

1. INVENTIONS AND INNOVATIONS: A LONG HISTORY AND MODERN INFATUATION

EVOLUTION AND HISTORY

American Society of Mechanical Engineers. 2022 "Owens AR Bottle Machine." Engineering History Landmarks no. 86. New York: ASME.

Librado, P., et al. 2021. "The Origins and Spread of Domestic Horses from the Western Eurasian Steppes." *Nature* 598:634–640.

Shea, J. J. 2016. *Stone Tools in Human Evolution: Behavioral Differences among Technological Primates*. Cambridge: Cambridge University Press.

Smil, V. 2018. *Energy and Civilization: A History*. Cambridge, MA: MIT Press.

Smil, V. 2014. *Making the Modern World: Materials and Dematerialization*. Chichester: Wiley.

INVENTION AND INNOVATION

Akana, J., et al. 2012. Portable display device. US Patent USD670,286S1, filed November 23, 2010, and issued November 6, 2012.

Brooks, D. E. 2013. Diane's manna. US Patent US8,609,158B2, filed June 20, 2012, and issued December 17, 2013.

Brown, A. E., and H. A. Jeffcott. 1932. *Beware of Imitations*. New York: Viking Press.

Carayannis, E. G. 2013. *Encyclopedia of Creativity, Invention, Innovation and Entrepreneurship*. Berlin: Springer.

Chan, C. L. 2015. "Fallen Behind: Science, Technology, and Soviet Statism." *Intersect: The Stanford Journal of Science, Technology, and Society* 8, no. 3.

Electronic Frontier Foundation. 2022. "Stupid Patent of the Month." Electronic Frontier Foundation. https://www.eff.org.

Hannas, W. C., and D. K. Tatlow., eds. 2021. *China's Quest for Foreign Technology: Beyond Espionage*. London: Routledge.

Perry, R. 1973. *Comparison of Soviet and US Technology*. Santa Monica, CA: Rand Corporation.

Sykes, A. O. 2021. "The Law and Economics of 'Forced' Technology Transfer and Its Implications for Trade and Investment Policy (and the U.S.-China Trade War)." *Journal of Legal Analysis* 13:127–171. https://doi.org/10.1093/jla/laaa007.

Tenner, E. 1997. *Why Things Bite Back: Technology and the Revenge of Unintended Consequences*. New York: Vintage.

Ever-faster?

Cannon, K. M., and D. T. Britt. 2019. "Feeding One Million People on Mars." *New Space* 7, no. 4 (December): 245–254.

Kurzweil, R. 2006. *The Singularity Is Near*. New York: Penguin.

Mokyr, J. 2014. "The Next Age of Invention: Technology's Future Is Brighter Than Pessimists Allow." *City Journal* 24 (Winter): 12–21. https://www.city-journal.org/html/next-age-invention-13618.html.

SpaceX. 2022. "Mars & Beyond: The Road to Making Humanity Multiplanetary." SpaceX.com. https://www.spacex.com/human-spaceflight/mars/.

US Patent and Trademark Office. 2021. U.S. Patent Activity Calendar Years 1790 to the Present (database). https://www.uspto.gov/web/offices/ac/ido/oeip/taf/h_counts.htm.

Failed designs

Cooper, G., and B. Sinclair. 1990. "Failed Innovations—ICOHTEC Symposium, Hamburg, August 1989." *Technology and Culture* 31:496–499.

Herring, S. D. 1989. *From the Titanic to the Challenger: An Annotated Bibliography on Technological Failures of the Twentieth Century*. New York: Garland Press.

Petroski, H. 1985. *To Engineer Is Human: The Role of Failure in Successful Design*. New York: St. Martin's Press.

Petroski, H. 2001. "The Success of Failure." *Technology and Culture* 42:321–328.

Schiffer, M. B. 2019. *Spectacular Flops: Game-Changing Technologies That Failed*. Clinton Corners, NY: Eliot Werner Publications.

Tracy, P. 2022. "Apple's 12 Most Embarrassing Product Failures." https://gizmodo.com/apple-failures-newton-pippin-butterfly-keyboard-macinto-1849106570.

Real world

Centers for Disease Control and Prevention. 2020. "Road Traffic Injuries and Deaths—A Global Problem." CDC, National Center for Injury Prevention and Control (last reviewed December 14).

McNish, J., and S. Silcoff. 2015. *Losing the Signal: The Untold Story behind the Extraordinary Rise and Spectacular Fall of BlackBerry*. New York: Flatiron Books.

Newall, P. 2018. *Ocean Liners: An Illustrated History*. Barnsley: Seaforth Publishing.

Smil, V. 2016. *Still the Iron Age: Iron and Steel in the Modern World*. Oxford: Elsevier.

2. INVENTIONS THAT TURNED FROM WELCOME TO UNDESIRABLE

LEADED GASOLINE

Engine knock

Lounici, M. S., et al. 2017. "Knock Characterization and Development of a New Knock Indicator for Dual-Fuel Engines." *Energy* 141, 2351e2361.

Zhen, X., et al. 2012. "The Engine Knock Analysis—An Overview." *Applied Energy* 92:628–636.

Octane numbers

Anderson, J. E., et al. 2012. "Octane Numbers of Ethanol-Gasoline Blends: Measurements and Novel Estimation Method from Molar Composition." SAE Technical Paper 2012-01-1274. doi: 10.4271/2012-01-1274.

Stolark, J. 2016. "Fact Sheet: A Brief History of Octane in Gasoline: From Lead to Ethanol." White Paper. Washington, DC: Environmental and Energy Study Institute.

History of leaded gasoline

Boyd, T. A. 2002. *Charles F. Kettering: A Biography*. Fairless Hills, PA: Beard Books.

Hagner, C. 1999. *Historical Review of European Gasoline Lead Content Regulations and Their Impact on German Industrial Markets*. Geesthacht: GKSS-Forschungszentrum Geesthacht GmbH.

Landrigan, P. J. 2002. "The Worldwide Problem of Lead in Petrol." *Bulletin of the World Health Organization* 80:768.

Midgley, T. IV. 2001. *From the Periodic Table to Production: The Life of Thomas Midgley, Jr., the Inventor of Ethyl Gasoline and Freon Refrigerants*. Corona, CA: Stargazer Publishing.

Nriagu, J. O. 1990. "Rise and Fall of Leaded Gasoline." *The Science of the Total Environment* 92:13–28.

Robert, J. C. 1983. *Ethyl—A History of the Corporation and the People Who Made It*. Charlottesville: University of Virginia Press.

Controversy over leaded gasoline during the 1920s

Denworth, L. 2009. *Toxic Truth: A Scientist, a Doctor, and the Battle over Lead*. Boston: Beacon Press.

Kovarik, W. 2003. "Ethyl: The 1920s Conflict over Leaded Gasoline and Alternative Fuels." Personal website of Prof. Kovarik. billkovarik.com.

Kovarik, W. 2005. "Milestones: Leaded Gasoline." *International Journal of Occupational and Environmental Health* 11:384–397.

Rosner, D., and G. Markowitz. 1985. "A 'Gift of God'? The Public Health Controversy over Leaded Gasoline during the 1920s." *American Journal of Public Health* 75:344–352.

Sicherman, B. 1984. *Alice Hamilton: A Life in Letters*. Cambridge, MA: Harvard University Press.

Phasing out leaded gasoline

Newell, R. G., and K. Rogers. 2003. "The U.S. Experience with the Phasedown of Lead in Gasoline." Discussion Paper. Washington, DC: Resources for the Future. https://web.mit.edu/ckolstad/www/Newell.pdf.

Nielsen, C. 2021. *Unleaded: How Changing Our Gasoline Changed Everything*. New Brunswick, NJ: Rutgers University Press.

US EPA (Environmental Protection Agency). 1985. *Costs and Benefits of Reducing Lead in Gasoline: Final Regulatory Impact Analysis*. Washington, DC: Office of Policy Analysis.

Lead poisoning

Lewis, J. 1985. "Lead Poisoning: A Historical Perspective." *EPA Journal* 11, no. 4 (May): 15–18.

Needleman, H. L. 1999. "History of Lead Poisoning in the World." Tucson, AZ: Center for Biological Diversity.

Riva, M. A., et al. 2012. "Lead Poisoning." *Safety and Health at Work* 3:11–16.

Lead neurotoxicity in children

Aizer, A., et al. 2016. "Do Low Levels of Blood Lead Reduce Children's Future Test Scores?" NBER Working Paper 2258. Cambridge, MA: National Bureau of Economic Research.

Bellinger, D. C., 2011. "The Protean Toxicities of Lead: New Chapters in a Familiar Story." *International Journal of Environmental Research and Public Health* 8:2593–2628.

Canfield, R. L., et al. 2004. "Impaired Neuropsychological Functioning in Lead-Exposed Children." *Developmental Neuropsychology* 26:513–540.

Markowitz, G., and D. Rosner. 2014. *Lead Wars: The Politics of Science and the Fate of America's Children*. Berkeley: University of California Press.

Mason, L. H., et al. 2014. "Pb Neurotoxicity: Neuropsychological Effects of Lead Toxicity." *Biomedical Research International*, article ID 840547.

Nwobi, N. L., et al. 2019. "Positive and Inverse Correlation of Blood Lead Level with Erythrocyte Acetylcholinesterase and Intelligence Quotient in Children: Implications for Neurotoxicity." *Interdisciplinary Toxicology* 12:136–142.

DDT

Discovery of DDT, its properties and benefits

IPCS INCHEM. 1999. *DDT*. Poisons Information Monograph 127. https://inchem.org/documents/pims/chemical/pim127.htm.

Müller, P. H. 1948. "Dichloro-Diphenyl-Trichloroethane and Newer Insecticides." Nobel Lecture, December 11, 1948. https://www.nobelprize.org/uploads/2018/06/muller-lecture.pdf.

Müller, P. H. 1961. "Zwanzig Jahre wissenschaftliche-synthetische Bearbeitung des Gebietes der synthetischen Insektizide." *Naturwissenschaftliche Rundschau* 14:209–219.

National Academy of Sciences, Committee on Research in the Life Sciences. 1970. *The Life Sciences*. Washington, DC: National Academy of Sciences.

Silent Spring

Carson, R. 1962. *Silent Spring*. Boston: Houghton and Mifflin.

Culver, L., et al., eds. 2012. *Rachel Carson's Silent Spring Encounters and Legacies*. Munich: Rachel Carson Center.

Dunlap, T. R., ed. 2008. *DDT, Silent Spring, and the Rise of Environmentalism*. Seattle: University of Washington Press.

Jameson, C. M. 2013. *Silent Spring Revisited*. London: A&C Black.

Kroll, G. 2001. "The 'Silent Springs' of Rachel Carson: Mass Media and the Origins of Modern Environmentalism." *Public Understanding of Science* 10:403–420.

US DDT ban

Ruckelshaus, W. 1972. "Consolidated DDT Hearing: Opinion and Order of the Administrator." *Federal Register* 37:13369–13376.

Secretary of State. 2016. "Bill Ruckelshaus: The Conscience of 'Mr. Clean.'" Legacy Washington.

Sweeney, E. M. 1972. "Hearing Examiner's Recommended Findings, Conclusions, and Orders." *Federal Register,* April 25, 40 CFR 164.32.

Whitney, C. 2012. "The Silent Decade: Why It Took Ten Years to Ban DDT in the United States." *Virginia Tech Undergraduate Historical Review* 1. http://doi.org/10.21061/vtuhr.v1i0.5.

Eggshell thinning

Barker, R. J. 1958. "Notes on Some Ecological Effects of DDT Sprayed on Elms." *Journal of Wildlife Management* 22:269–274.

Falk, K., et al. 2018. "Raptors Are Still Affected by Environmental Pollutants: Greenlandic Peregrines Will Not Have Normal Eggshell Thickness until 2034." *Ornis Hungarica* 26:171–176.

Peakall, D. B. 1993. "DDE-Induced Eggshell Thinning: An Environmental Detective Story." *Environmental Review* 1:13–20.

Ratcliffe, D. A. 1958. "Broken Eggs in Peregrine Eyries." *British Birds* 51:23–26.

Ratcliffe, D. A. 1967. "Decrease in Eggshell Weight in Certain Birds of Prey." *Nature* 215:208–210.

DDT and malaria

Bouwman, H., et al. 2011. "DDT and Malaria Prevention: Addressing the Paradox." *Environmental Health Perspectives* 119:744–747.

Buxton, P. A. 1945. "The Use of the New Insecticide DDT in Relation to the Problems of Tropical Medicine." *Transactions of the Royal Society of Tropical Medicine and Hygiene* 38:367–400. https://doi.org/10.1016/0035-9203(45)90039-3.

Dagen, M. 2020. "History of Malaria and Its Treatment." In G. L. Patrick, ed., *Antimalarial Agents*, Amsterdam: Elsevier, 1–48.

Palmer, M. 2016. "The Ban of DDT Did Not Cause Millions to Die from Malaria." https://www.semanticscholar.org/paper/The-ban-of-DDT-did-not-cause-millions-to-die-from-Palmer/0e6812f87d27be92effac4fe8bfd414bc8f82476.

Pruett, B. D. 2013. "Dichlorophenyltrichloroethane (DDT): A Weapon Missing from the U.S. Department of Defense's Vector Control Arsenal." *Military Medicine* 178:243–245.

UN Environment Program. 2001. *Stockholm Convention on Persistent Organic Pollutants*. New York: UNEP.

UN Environment Program. 2010. *Ridding the World of POPs: A Guide to the Stockholm Convention on Persistent Organic Pollutants*. Geneva: Stockholm Convention Secretariat, UNEP. http://chm.pops.int/Portals/0/Repository/CHM-general/UNEP-POPS-CHM-GUID-RIDDING.English.PDF.

World Health Organization. 2011. *The Use of DDT in Malaria Vector Control*. Geneva: WHO.

World Health Organization. 2020. *World Malaria Report 2020: 20 Years of Global Progress and Challenges*. Geneva: WHO.

DDT and human health

Eskenazi, B., et al. 2009. "The Pine River Statement: Human Health Consequences of DDT Use." *Environmental Health Perspectives* 117:1359–1367.

Larsen, N. 2021. "Thomas Midgley, the Most Harmful Inventor in History." Podcast. https://www.bbvaopenmind.com/en/science/research/thomas-midgley-harmful-inventor-history.

Rogan, W. J. and A. Chen. 2005. "Health Risks and Benefits of Bi(4-Chlorophenyl)-1,1,1-Trichloroethane (DDT)." *Lancet* 366:763–773.

US Department of Health and Human Services. 2019. "Toxicological Profile for DDT, DDE, and DDD." Washington, DC: USDHHS.

CHLOROFLUOROCARBONS

CFCs

Calm. J. M. 2008. "The Next Generation of Refrigerants: Historical Review, Considerations, and Outlook." *International Journal of Refrigeration* 31:1123–1133.

Giunta, C. J. 2006. "Thomas Midgley, Jr., and the Invention of Chlorofluorocarbon Refrigerants: It Ain't Necessarily So." *Bulletin for the History of Chemistry* 31:66–74.

McLinden, M. O., and M. L. Huber. 2020. "(R)Evolution of Refrigerants." *Journal of Chemical & Engineering Data* 65:4176–4193.

Midgley, T. Jr. 1937. "From the Periodic Table to Production." *Industrial and Engineering Chemistry* 29:241–244.

Midgley, T. Jr., and A. L. Henne. 1930. "Organic Fluorides as Refrigerants." *Industrial and Engineering Chemistry* 22:542–545.

Midgley, T. Jr., A. L. Henne, and R. R. McNary. 1931. Heat transfer. US Patent 1,833,847, issued November 24, 1931.

Midgley, T. IV. 2001. *From the Periodic Table to Production: The Life of Thomas Midgley, Jr., the Inventor of Ethyl Gasoline and Freon Refrigerants.* Corona, CA: Stargazer Publishing.

Rigby, M., et al. 2013. "Re-evaluation of the Lifetimes of the Major CFCs and CH_3CCl_3 Using Atmospheric Trends." *Atmospheric Chemistry and Physics* 13:2691–2702.

Sicard, A. J., and R. T. Baker. 2020. "Fluorocarbon Refrigerants and Their Syntheses: Past to Present." *Chemical Reviews* 120:9164–9303.

CFCs and ozone

Cagin, S., and P. Dray. 1993. *Between Earth and Sky: How CFCs Changed Our World and Endangered the Ozone Layer.* New York: Pantheon.

Dotto, L., and H. Schiff. 1978. *The Ozone War.* Garden City, NY: Doubleday & Co.

Douglass, A. R., et al. 2014. "The Antarctic Ozone Hole: An Update." *Physics Today* 67, no. 7 (July): 42. doi: 10.1063/PT.3.2449.

Farman, J. C., et al. 1985. "Large losses of Total Ozone in Antarctica Reveal Seasonal ClO_x/NO_x Interaction." *Nature* 315:207–210.

Lovelock, J. E. 1971. "Atmospheric Fluorine Compounds as Indicators of Air Movements." *Nature* 230:379.

Molina, M. J. 1995. "Polar Ozone Depletion." Nobel Lecture, December 8, 1995. https://www.nobelprize.org/uploads/2018/06/molina-lecture.pdf.

Molina, M. J., and F. S. Rowland. 1974. "Stratospheric Sink for Chlorofluoromethanes: Chlorine Atom Catalyzed Destruction of Ozone." *Nature* 249:810–812.

NASA. 2019. "Ozone Hole Is the Smallest on Record Since Its Discovery." NASA, October 21. https://www.nasa.gov/feature/goddard/2019/2019-ozone-hole-is-the-smallest-on-record-since-its-discovery.

NASA. 2020. "Large, Deep Antarctic Ozone Hole in 2020." NASA Earth Observatory, September 20. https://earthobservatory.nasa.gov/images/147465/large-deep-antarctic-ozone-hole-in-2020.

Newman, P.A., et al. 2009. "What Would Have Happened to the Ozone Layer If Chlorofluorocarbons (CFCs) Had Not Been Regulated?" *Atmospheric Chemistry and Physics* 9:2113–2128.

Roan, S. 1989. *Ozone Crisis: The 15-year Evolution of a Sudden Global Emergency*. New York: John Wiley & Sons.

Rowland, F. S. 1995. Nobel Lecture in Chemistry. December 8. https://www.nobelprize.org/uploads/2018/06/rowland-lecture.pdf.

Solomon, S. 1999. "Stratospheric Ozone Depletion: Review of Concepts and History." *Review of Geophysics* 37:375–316.

Stolarski, R. S., and R. J. Cicerone. 1974. "Stratospheric Chlorine: A Possible Sink for Ozone." *Canadian Journal of Chemistry* 52:1610–1615.

Tevini, M., ed. 1993. *UV-B Radiation and Ozone Depletion: Effects on Humans, Animals, Plants, Microorganisms, and Materials*. Boca Raton, FL: Lewis Publishers.

CFC ban and refrigerant replacements

Maxwell, J., and F. Briscoe. 1997. "There's Money in the Air: The CGC Ban and DuPont's Regulatory Strategy." *Business Strategy and the Environment* 6:276–286.

Reimann, C. R. 2018. "Observing the Atmospheric Evolution of Ozone-Depleting Substances." *Geoscience* 350:384–392.

United Nations Environment Program. 2020. *Montreal Protocol on Substances That Deplete the Ozone Layer*. Nairobi: UNEP.

3. INVENTIONS THAT WERE TO DOMINATE—AND DO NOT

AIRSHIPS

History

Dwiggins, D. 1980. *The Complete Book of Airships—Dirigibles, Blimps and Hot Air Balloons*. Shrewsbury: Airlife.

Folkes, J. 2008. "Balloons, Airships and Kites—Lighter Than Air: Past, Present and Future." *Aeronautical Journal* 112:421–429.

Liao, L., and I. Pasternak. 2009. "A Review of Airship Structural Research and Development." *Progress in Aerospace* 45:83–96.

MacMechen, T. R., and C. Dienstbach. 1912. "The Greyhounds of the Air." *Everybody's Magazine* 27:290–304.

Robinson, D. H. 1973. *Giants in the Sky: History of the Rigid Airship*. Henley-on-Thames: Foulis.

Swinfield, J. 2012. *Airship: Design, Development and Disaster*. London: Conway.

Toland, J. 1972. *The Great Dirigibles: Their Triumphs and Disasters*. Mineola, NY: Dover Publishers, 1972.

Airships in the military

Dienstbach, C., and T. R. MacMechen. 1909. "The Aerial Battleship." *McClure's Magazine* 33:422–434.

Jamison, L., et al. 2005. *High-Altitude Airships for the Future Force Army*. Santa Monica, CA: Rand Corporation.

Robinson, D. W. 1976. *USAF History of Manned Balloons and Airships*. Maxwell Air Force Base, AL: USAF.

Robinson, D. W. 1997. *The Zeppelin in Combat: A History of the German Naval Airship Division, 1912–1918*. Seattle: University of Washington Press.

Zeppelins

Archbold, R., and K. Marschall. 1994. *Hindenburg: An Illustrated History*. New York: Warner Books.

Botting, D. 2001 *Dr. Eckener's Dream Machine: The Great Zeppelin and the Dawn of Air Travel*. New York: Henry Holt and Co.

Brooks, P. 2004. *Zeppelin: Rigid Airships 1893–1940*. London: Putnam Aeronautical Books.

Clausberg, K. 1979. *Zeppelin: Die Geschichte eines unwahrscheinlichen Erfolges*. München: Schirmer/Mosel.

de Syon, G. 2001. *Zeppelin! Germany and the Airship, 1900–1939*. Baltimore, MD: Johns Hopkins University Press.

Dick, H. G., and D. H. Robinson. 1985 *The Golden Age of the Great Passenger Airships Graf Zeppelin & Hindenburg*. Washington, DC: Smithsonian Institution Press.

DiLisi, G. A. 2017. "The Hindenburg Disaster: Combining Physics and History in the Laboratory." *Physics Teacher* 55:268.

Eckener, H. 1929. Rigid airship with separate gas cells. US Patent 1,724,009, issued August 13, 1929.

Eckener, H. 1958. *My Zeppelins*. London: Putnam and Co.

Lehmann, E. 1937. *Zeppelin: The Story of Lighter-Than-Air Craft*. London: Longmans, Green and Co.

Majoor, M. 2000. *Inside the Hindenburg*. Boston: Little, Brown and Co.

Sattelmacher, A. 2021. "Shuffled Zeppelin Clips: The Flight and Crash of LZ 129 Hindenburg in the Archives." *Isis* 112:352–360.

Zeppelin, F. 1899. Navigable balloon. US Patent 1,621,195, filed December 29, 1897, and issued March 14, 1899.

Future

Ling, J. 2020. "The Age of the Airship May Be Dawning Again." *Foreign Policy,* February 29.

Miller, S., et al. 2014. *Airships: A New Horizon for Science*. Pasadena, CA: Keck Institute for Space Studies. www.kiss.caltech.edu/study/airship/.

Prentice, B. E., et al. 2021. "Transport Airships for Scheduled Supply and Emergency Response in the Arctic." *Sustainability* 13:5301.

Windischbauer, F., and J. Richardson. 2005. "Is There Another Chance for Lighter-Than-Air Vehicles?" *Foresight* 7:54–65.

Villamizar, H. 2022. "Air Nostrum Orders a Fleet of Airlander Airships." *Airways Magazine,* June 17, 2022. https://airwaysmag.com/air-nostrum-airlander-airships.

NUCLEAR FISSION

The road to nuclear fission

Hahn, O., and F. Strassman. 1939. "Über den Nachweis und das Verhalten der bei der Bestrahlung des Urans mittles Neutronen entstehenden Erdalkalimetalle." *Naturwissenschaften* 27:11–15.

Lanouette, W. 1992. *Genius in Shadows: A Biography of Leo Szilard*. New York: Charles Scribner's Sons.

Meitner, L., and O. R. Frisch. 1939. "Disintegration of Uranium by Neutrons: A New Type of Nuclear Reaction." *Nature* 143:239–240.

History of nuclear electricity generation

Eisenhower, D. D. 1953. Atoms for Peace Speech to the 470th Plenary Meeting of the United Nations General Assembly. https://www.iaea.org/about/history/atoms-for-peace-speech.

Beck, P. W. 1999. "Nuclear Energy in the Twenty-First Century: Examination of a Contentious Subject." *Annual Review of Energy* 24:113–138.

Lovering, J. R., et al. 2016. "Historical Construction Costs of Global Nuclear Power Reactors." *Energy Policy* 91:371–382.

Marcus, G. 2010. *Nuclear Firsts: Milestone on the Road to Nuclear Power Development.* La Grange Park, IL: American Nuclear Society.

Murray, R. L. 2009. *Nuclear Energy*. Oxford: Elsevier.

Nuclear generation in the US

Cantelon, P. L. 1984. *The American Atom: A Documentary History of Nuclear Policies from the Discovery of Fission to the Present, 1939–1984*. Philadelphia: University of Philadelphia Press.

Cowan, R. 1990. "Nuclear Power Reactors: A Study in Technological Lock-in." *Journal of Economic History* 50:541–567.

Forsberg, C. W., and A. M. Weinberg. 1990. "Advanced Reactors, Passive Safety, and Acceptance of Nuclear Energy." *Annual Review of Energy* 15:133–152.

Holl, J. M., et al. 1985. *United States Civilian Nuclear Power Policy, 1954–1984: A History*. Washington, DC: US Department of Energy.

Kaplan, S. 2008. *Power Plants: Characteristics and Costs*. Washington, DC: Congressional Research Service.

Lowen, R. S. 1987. "Entering the Atomic Power Race: Science, Industry, and Government." *Political Science Quarterly* 102:459–479.

Nuclear Regulatory Commission. 2011. *Reactor Designs, Safety, Emergency Preparedness, Security, Renewals, New Designs, Licensing, American Plants, Decommissioning.* Washington, DC: NRC.

Parker, L., and M. Holt. 2007. *Nuclear Power: Outlook for New U.S. Reactors*. Washington, DC: Congressional Research Service.

Pope, D. 1991. "Seduced and Abandoned? Utilities and WPPSS Nuclear Plants 4 and 5." *Columbia Magazine*, Fall 1991.

Rockwell, T. 1992. *The Rickover Effect: How One Man Made a Difference*. Annapolis, MD: Naval Institute Press.

Nuclear risks and accidents

Chapin, D. M., et al. 2002. "Nuclear Power Plants and Their Fuel as Terrorist Target." *Science* 297:1997–1999.

Mahaffey, J. 2019. *Atomic Accidents: A History of Nuclear Meltdowns and Disasters: From the Ozark Mountains to Fukushima*. Oakland, CA: Pegasus Books.

Nuclear Energy Agency. 2002. *Chernobyl: Assessment of Radiological and Health Impacts*. Paris: NEA.

Weinberg, Alvin M. 1972. "Social Institutions and Nuclear Energy." *Science* 177: 27–34.

Fast breeder reactors

Cochran, T. B., et al. 2010. *Fast Breeder Reactor Programs: History and Status*. Princeton, NJ: International Panel on Fissile Materials.

International Atomic Energy Agency. 2012. *Status of Fast Reactor Research and Technology Development*. Vienna: IAEA. https://www-pub.iaea.org/MTCD/Publications/PDF/te_1691_web.pdf.

Judd, A. M. 1981. *Fast Breeder Reactors: An Engineering Introduction*. Oxford: Pergamon.

Sokolski, H. 2019. "The Rise and Demise of the Clinch River Breeder Reactor." *Bulletin of the Atomic Scientists*, February 6. https://thebulletin.org/2019/02/the-rise-and-demise-of-the-clinch-river-breeder-reactor/.

Small modular reactors

IAEA. 2021. Small nuclear power reactors.

Oklo. 2021. "What Could You Do with a MW-Decade of Emission-Free Power?," https://oklo.com.

Rolls Royce. 2021. "Small Modular Reactors—Rolls-Royce." https://www.rolls-royce.com/innovation/small-modular-reactors.aspx.

TerraPower. 2021. The Natrium Reactor: From Research to Reality. https://www.terrapower.com/natrium-reactor-reality-2021.

World Nuclear Association. "Small Nuclear Power Reactors." http://world-nuclear.org.

SUPERSONIC FLIGHT

History

Bisplinghoff, R. L. 1964. "The Supersonic Transport." *Scientific American* 210, no. 6: 25–35.

Culick, F. E. C. 1979. "The Origins of the First Powered, Man-Carrying Airplane." *Scientific American* 241, no. 1: 86–100.

International Civil Aviation Organization. 1960. *Annual Report of the Council to the Assembly for 1959*. Montreal: ICAO. https://www.icao.int/assembly-archive/Session 13E/A.13.REP.1.P.EN.pdf.

General considerations

Carioscia, S. A., et al. 2019. *Challenges to Supersonic Flight*. Alexandria, VA: Institute for Defense Analyses.

Edwards, G. 1974. "The Technical Aspects of Supersonic Civil Transport Aircraft." *Philosophical Transactions of the Royal Society of London. Series A, Mathematical and Physical Sciences* 275:529–565.

Nowlan, F. S., and K. W. Comstock. 1965. "The Assessment of Supersonic Transport Operating Costs." *SAE Transactions* 73:685–697.

Tang, R. Y., et al. 2018. *Supersonic Passenger Flights*. Washington, DC: Congressional Research Service.

Concorde

Bureau d'Enquêtes et d'Analyses pour le Sécurité de l'Aviation Civil. 2002. *Accident on 25 July 2000 at La Patte d'Oie in Gonesse (95) to the Concorde Registered F-BTSC Operated by Air France*. Paris: BEA.

Butcher, L. 2010. *Aviation: Concorde*. London: Library House of Commons.

Buttler, T., and J. Carbonel. 2018. *Building Concorde: From Drawing Board to Mach 2*. Forest Lake, MN: Specialty Press.

Glancy, J. 2016. *Concorde: The Rise and Fall of the Supersonic Airliner*. Boston: Atlantic Books.

Johnman, L., and F. M. B. Lynch. 2002. "The Road to Concorde: Franco-British Relations and the Supersonic Project." *Contemporary European History* 11:229–252.

Smith, R. K. 2019. "The Supersonic Airliner Fiasco: Frenzied International Aeronautical Saga of Communicable Obsessions, 1956–1976." *Air Power History*, Fall, 5–20.

Trubshaw, B. 2019. *Concorde: The Complete Inside Story*. Cheltenham: History Press.

American SST

Bedwell, D. 2012. "Supersonic Gamble." *Aviation History Magazine*, May. https://www.historynet.com/supersonic-gamble.htm.

Office of Technology Assessment. 1980. *Impact of Advanced Air Transport Technology*. Washington, DC: OTA.

Recent projects

Boom Supersonic. 2022. "Boom—Supersonic Passenger Airplanes." https://boomsupersonic.com/.

Lockheed Martin. 2022. "X-59." https://www.lockheedmartin.com/en-us/products/quesst.html.

Schneider, D. 2021. "The Recent Supersonic Boom. *Spectrum IEEE,* August.

Spike Aerospace. 2022. "The Spike S-512 Supersonic Jet: Fly Supersonic. Do More." https://www.spikeaerospace.com/.

4. INVENTIONS THAT WE KEEP WAITING FOR

TRAVEL IN A (NEAR) VACUUM (HYPERLOOP)

George Medhurst

London Mechanics' Register 1825. London and Edinburgh Vacuum Tunnel Company. 1825. *London Mechanics' Register* 1:205-207.

Medhurst, G. 1812. *Calculations and Remarks, Tending to Prove the Practicality, Effects and Advantages of a Plan for the Rapid Conveyance of Goods and Passengers Upon an Iron Road Through a Tube of 30 Feet in Area, by the Power and Velocity of Air.* London: D. N. Shury.

Medhurst, G. 1827. *A New System of Inland Conveyance, for Goods and Passengers, Capable of Being Applied and Extended Throughout the Country; and of Conveying All Kinds of Goods, Cattle, and Passengers, with the Velocity of Sixty Miles in an Hour, at an Expense That Will Not Exceed the One-Fourth Part of the Present Mode of Travelling, Without the Aid of Horses or Any Animal Power.* London: T. Brettell.

Isambard K. Brunel

Buchanan, R. A. 1992. "The Atmospheric Railway of I. K. Brunel." *Social Studies of Science* 22:231–243.

Robert H. Goddard

Goddard, R. H. 1914. "Bachelet's Frictionless Railway at Basis a Tech Idea." *Worcester Polytechnic Institute Journal* 1914:12–21.

Goddard, R. H. 1950. Vacuum tube transportation system. US Patent 2,511,979A, filed May 21, 1945, issued June 30,1950.

Scientific American. 1909. "The Limit of Rapid Transit." *Scientific American* 101, no. 1: 366.

Émile Bachelet

Bachelet, É. 1912. Levitating transmitting apparatus. US Patent 1,020,942, issued March 19, 1912.

B. P. Weinberg

Weinberg, B. 1917. "Traveling at 500 Miles Per Hour in the Future Electric Railway." *Electrical Experimenter* 1917:794.

Weinberg, B. 1919. "Traveling at 500 Miles an Hour." *Popular Science Monthly* 1919:705.

R. B. Davy

Davy, R. B. 1920. Vacuum-railway. US Patent 1,336,782, issued April 13, 2020.

Robert Salter

Salter, R. M. 1972. *The Very High Speed Transit System*. Santa Monica, CA: Rand Corporation.

Salter, R. M. 1978. *Trans-Planetary Subway Systems—A Burgeoning Capability*. Santa Monica, CA: Rand Corporation.

Hyperloop

Armagana, K. 2020. "The Fifth Mode of Transportation: Hyperloop." *Journal of Innovative Transportation* 1, no. 1: 1105.

Hyperloop TT. 2022. "The Future Is Now Boarding." Hyperloop Transportation Technologies. https://www.hyperlooptt.com.

Klühspies, J. et al. 2022. *Hyperloop? Ergebnisse einer internationalen Umfrage im Verkehrswesen*. Munich: The International Maglev Board.

Musk, E. 2013. "Hyperloop Alpha." https://www.tesla.com/sites/default/files/blog_images/hyperloop-alpha.pdf.

Nøland, K. 2021. "Prospects and Challenges of the Hyperloop Transportation System: A Systematic Technology Review." *IEEE Access* 9:28439–28458. https://ieeexplore.ieee.org/stamp/stamp.jsp?arnumber=9350309.

Virgin Hyperloop. 2021. "Virgin Hyperloop." https://virginhyperloop.com.

NITROGEN-FIXING CEREALS

History of diazotrophs

Beijerinck, M. W. 1888. Die Bakterien der Papilionaceen-Knölchen. *Botanische Zeitschrift* 46:725–804.

Boddey, R. M., and J. Döbereiner. 1995. "Nitrogen Fixation Associated with Grasses and Cereals: Recent Progress and Perspectives for the Future." *Fertilizer Research* 42:241–250.

Borlaug, N. 1970. "Nobel Prize Acceptance Speech." December 10. https://www.nobelprize.org/prizes/peace/1970/borlaug/acceptance-speech/.

Burrill, T. J., and R. Hansen. 1917. "Is Symbiosis Possible between Legume Bacteria and Non-Legume Plants?" *Agricultural Experimental Station Bulletin* 202:115–181.

Döbereiner, J. 1988. "Isolation and Identification of Root Associated Diazotrophs." *Plant and Soil* 110:207–212.

Hellriegel, H., and H. Wilfarth. 1888. "Untersuchungen über die Stickstoffernährung der Gramineen und Leguminosen." *Beiläge der Zeitschrift des Vereins für die Rübenzuckerindustrie*. Berlin: Kayssler.

Löhnis, F. 1921. "Nodule Bacteria of Leguminous Plants." *Journal of Agricultural Research* 20:543–556.

Smil, V. 2001. *Enriching the Earth: Fritz Haber, Carl Bosch and the Transformation of World Food Production*. Cambridge, MA: MIT Press.

Fixation in cereals

Beatty, P. H., and A. G. Good. 2011. "Future Prospects for Cereals That Fix Nitrogen." *Science* 333:416–417.

Bloch, S. E. et al. 2020. "Harnessing Atmospheric Nitrogen for Cereal Crop Production." *Current Opinion in Biotechnology* 62:181–188.

Crookes, W. 1898. "Address of the President before the British Association for the Advancement of Science, Bristol, 1898." *Science* 8:561–575.

Huisman, R., and R. Geurts. 2020. "A Roadmap toward Engineered Nitrogen Fixing Nodule Symbiosis." *Plant Communications*. https://doi.org/10.1016/j.xplc.2019.100019

Pankiewicz, V. C. S., et al. 2019. "Are We There Yet? The Long Walk towards the Development of Efficient Symbiotic Associations between Nitrogen-Fixing Bacteria and Non-Leguminous Crops." *BMC Biology* 17:99.

Rosenblueth, M., et al. 2018. "Nitrogen Fixation in Cereals." *Frontiers in Microbiology* 9:1794. doi: 10.3389/fmicb.2018.0179.

Sharma, P., et al. 2016. "Biological Nitrogen Fixation in Cereals: An Overview." *Journal of Wheat Research* 8, no. 2: 1–11.

Yang, J., et al. 2018. "Polyprotein Strategy for Stoichiometric Assembly of Nitrogen Fixation Components for Synthetic Biology." *Proceedings of the National Academy of Sciences* 115, no. 36: E8509-E8517.

Recent news

Azotic Technologies. 2018. "Azotic's Natural Nitrogen Fixing Technology Is Now Commercially Available in the USA." https://www.azotictechnologies.com/news-and-insight/latest-news/heading-5/#:~:text=After%20positive%20field%20trial%20results,results%20and%20feedback%20from%20growers.

Azotic Technologies. 2021. "Envita Technologies." https://www.azotic-na.com/science-behind-envita.

Schwartz, J., et al. 2020. *Practical Farm Research 2020*. https://www.beckshybrids.com/portals/0/sitecontent/literature/2020-2021-literature/Becks-2020-PFRBook.pdf.

Schwartz, J., et al. 2021, *Practical Farm Research 2021*. https://www.beckshybrids.com/portals/0/sitecontent/literature/2021-2022-literature/PFR-Book-2021-web.pdf.

US Food and Drug Administration. 2021. *GMO Crops, Animal Food, and Beyond*. Washington, DC: US FDA.

Witt, M., et al. 2020. *On-Farm Corn Nitrogen Enhancer Foliar Treatment Demonstration Trials*. Ames: Iowa State University.

CONTROLLED NUCLEAR FUSION

Sun

Bethe, H. A. 1967. "Energy Production in Stars." Nobel Lecture, December 11. https://www.nobelprize.org/uploads/2018/06/bethe-lecture.pdf.

Physics of fusion

Glasstone, S. 1974. *Controlled Nuclear Fusion*. Washington, DC: US Atomic Energy Commission.

Kikuchi, M., et al., eds. 2012. *Fusion Physics*. Vienna: International Atomic Energy Agency.

History of fusion research

Chou, C. B., et al. 2016. *Fusion Energy via Magnetic Confinement: An Energy Technology Distillate*. Princeton, NJ: Andlinger Center for Energy and the Environment.

Coppi, B. 2016. "Relevance of Advanced Nuclear Fusion Research: Breakthroughs and Obstructions." *American Institute of Physics Conference Proceedings* 1721, no. 1, 020003. https://doi.org/10.1063/1.4944012.

Dean, S. O. 2013. *Search for the Ultimate Energy Source: A History of the U.S. Fusion Energy Program*. New York: Springer.

El-Guebaly, L. 2010. "Fifty Years of Magnetic Fusion Research (1958–2008): Brief Historical Overview and Discussion of Future Trends." *Energies* 3:1067–1086.

Lopes Cardozo, N. J., et al. 2016. "Fusion: Expensive and Taking Forever?" *Journal of Fusion Energy* 35:94–101

Shafranov, V. D. 2001. "On the History of the Research into Controlled Thermonuclear Fusion." *Uspekhi Fizicheskikh Nauk* 44:835–865.

Tokamaks

Zohm, H. 2019. "On the Size of Tokamak Fusion Power Plants." *Philosophical Transactions of the Royal Society A* 377: 20170437. http://dx.doi.org/10.1098/rsta.2017.043.

ITER

ITER. 2021. "What Is ITER?" https://www.iter.org/proj/inafewlines.

Inertial fusion

Nuckolls, J., et al. 1972. "Laser Compression of Matter to Super-High Densities: Thermonuclear (CTR) Applications." *Nature* 239:139–142.

Zylstra, A. B., et al. 2022. "Burning Plasma Achieved in Inertial Fusion." *Nature* 601:542–548.

Cold fusion (LENR)

Ball, P. 2019. "Lessons from Cold Fusion: 30 Years On." *Nature* 569:601.

Berlinguette, C. P., et al. 2019. "Revisiting the Cold Case of Cold Fusion." *Nature* 570:45–51.

Nagel, D. J. 2021. "Experimental Status of LENR." PowerPoint presentation. Washington, DC: US Department of Energy.

Future

Ball, P. 2021. "The Race to Fusion Energy." *Nature* 599:562–566.

Dabbar, P. 2021. "Fusion Breakthrough Dawns a New Era for US Energy and Industry. *The Hill,* September 10. https://thehill.com/opinion/technology/571722-fusion-breakthrough-dawns-a-new-era-for-us-energy-and-industry/.

Enter, S., et al. 2018. "Approximation of the Economy of Fusion Energy." *Energy* 152:489–497.

European Fusion Development Agreement. 2012. *Fusion Electricity: A Roadmap to the Realisation of Fusion Energy*. Culham: EFDA.

Galchen, R. "Green Dream." *New Yorker,* October 11, 22–28.

Hirsch, R. L. 2015. "Fusion Research: Time to Set a New Path." *Issues in Science and Technology* Summer 2015:35–42.

International Atomic Energy Agency. 2021. "Fusion Energy." *IAEA Bulletin,* May.

Jassby, D. 2017. "Fusion Reactors: Not What They're Cracked Up to Be." *Bulletin of the Atomic Scientists,* April 19.

Young, C. 2021. "We Are Now One Step Closer to Limitless Energy from Nuclear Fusion." *Interesting Engineering* September 9, 2021.

5. TECHNO-OPTIMISM, EXAGGERATIONS, AND REALISTIC EXPECTATIONS

BREAKTHROUGHS THAT ARE NOT

Gandy, S. 2021. "6 Ways the FDA's Approval of Aduhelm Does More Harm Than Good." *STAT,* June 15. https://www.statnews.com/2021/06/15/6-ways-fda-approval-aduhelm-does-more-harm-than-good/.

Hall, B. H. 2020. "Patents, Innovation, and Development." NBER Working Paper 27203. Cambridge, MA: National Bureau of Economic Research.

McDonald, L. 2017. "What Is Tony Seba Smoking? *EVAdoption News,* May 20. https://evadoption.com/what-is-tony-seba-smoking-evadoption-news-may-20-2017/.

Norris, M. 2020. "Brain-Computer Interfaces Are Coming. Will We Be Ready?" *The RAND blog*. Santa Monica, CA: Rand Corporation, August 27.

Pham, C., and F. Gilbert. 2021. "Predicting the Future of Brain-Computer Interface Technologies: The Risky Business of Irresponsible Speculation in News Media." *Bioethics Forum* 12:15–28.

RethinkX. 2017. *Transportation Report.* https://www.rethinkx.com/transportation.

Rosario, C. 2019. "4 Problems with Electronic Health Records." Advanced Data Systems Corporation, October 16. https://www.adsc.com/blog/problems-with-electronic-health-records.

SpaceX. 2017. "Mars & Beyond." https://www.spacex.com/human-spaceflight/mars/.

Sumner, P., et al. 2014. "The Association between Exaggeration in Health-Related Science News and Academic Press Releases: Retrospective Observational Study." *British Medical Journal* 2014:349. doi: 10.1136/bmj.g7015.

EVER-FASTER GROWTH?

Azhar, A. 2021. *The Exponential Age: How Accelerating Technology Is Transforming Business, Politics, and Society*. New York: Diversion Books.

Berlinski, D. 2018. "Godzooks." *Inference* 3, no. 4. https://inference-review.com/article/godzooks.

Harari, Y. 2017. *Homo Deus: A Brief History of Tomorrow*. New York: Harper.

Kurzweil, R. 2005. *The Singularity Is Near.* New York: Penguin.

Kurzweil, R. 2021. "Kurzweil: Tracking the Acceleration of Intelligence." http://www.kurzweilai.net/.

Mokyr, J. 2014. "The Next Age of Invention: Technology's Future Is Brighter Than Pessimists Allow." *City Journal* 24 (Winter): 12–21. https://www.city-journal.org/html/next-age-invention-13618.html.

Mokyr, J. 2017. *A Culture of Growth: The Origins of the Modern Economy*. Princeton, NJ: Princeton University Press.

DRUGS

Kinch, M. S. 2015. "An Overview of FDA-Approved Biologics Medicines." *Drug Discovery Today* 20:393–398.

Kinch, M. S., et al. 2013. "An Overview of FDA-Approved New Molecular Entities: 1827–2013." *Drug Discovery Today* 19:1033–1039.

Ricciarelli, R., and E. Fedele. 2017. "The Amyloid Cascade Hypothesis in Alzheimer's Disease: It's Time to Change Our Mind." *Current Neuropharmacology* 15:926–935.

US Food and Drug Administration. 2022. Drug Approvals and Databases.

AVIATION

Ahlgren, L. 2021. "Embraer Launches a Fleet of 4 New Sustainable Aircraft Designs." Simple Flying, November 8. https://simpleflying.com/embraer-sustainable-aircraft-designs/.

Bailey, J. 2019. "Who Is Alice? An Introduction to the Bizarre Eviation Electric Aircraft." Simple Flying, June 26. https://simpleflying.com/eviation-alice-electric-aircraft.

Eviation. 2022. "Sustainable, Economical Aviation." http://eviation.com.

Universal Hydrogen. 2021. "Fueling Carbon-Free Flight." https://hydrogen.aero/.

Zunum Aero. 2019. "Bringing You Electric Air Travel Out to a Thousand Miles." https://zunum.aero/.

ARTIFICIAL INTELLIGENCE

Anderson, J., et al. 2018. *Artificial Intelligence and the Future of Humans*. Washington, DC: Pew Research Center.

Choi, C. Q. 2021. "7 Revealing Ways AIs Fail." *IEEE Spectrum* September 2021:42–47.

Jordan, M. I. 2021. "Stop Calling Everything 'Artificial Intelligence.'" Mind Matters: News, April 7, 2021. https://mindmatters.ai/2021/04/ai-researcher-stop-calling-everything-artificial-intelligence/#:~:text=Jordan%20(pictured)%20adds%2C,talking%20as%20if%20we%20do.%E2%80%9D.

Kissinger, H., et al. 2021. *The Age of AI*. Boston: Little, Brown and Co.

Pretz, K. 2021. "Stop Calling Everything AI, Machine-Learning Pioneer Says." *IEEE Spectrum,* September, 58–59.

Roitblat, H. L. 2020. *Algorithms Are Not Enough: Creating General Artificial Intelligence*. Cambridge, MA: MIT Press.

Strickland, E. 2021. "The Turbulent Past and Uncertain Future of AI." *IEEE Spectrum,* October, 27–31.

Thompson, N. C., et al. 2021. "Deep Learning's Diminishing Returns." *IEEE Spectrum,* October, 51–55.

MOORE'S LAW

Hall, E. C. 1996. *Journey to the Moon: The History of the Apollo Guidance Computer.* Washington, DC: American Institute of Aeronautics and Astronautics.

Hennessy, J. 2019. The End of Moore's Law & Faster General Purpose Computing, and a Road Forward. Faculty paper, Stanford University, March. https://opennet working.org/wp-content/uploads/2020/12/9_2.05pm_John_Hennessey.pdf.

Moore, G. E. 1965. "Cramming More Components onto Integrated Circuits." *Electronics* 38, no. 8: 114–117.

Moore, G. E. 1975. "Progress in Digital Integrated Electronics." *Technical Digest, IEEE International Electron Devices Meeting,* 11–13.

Moore, G. E. 2003. "No Exponential Is Forever: But 'Forever' Can Be Delayed!" Paper presented at IEEE International Solid-State Circuits Conference, San Francisco. http://ieeexplore.ieee.org/document/1234194/.

Rupp, K., and S. Selberherr. 2011. "The Economic Limit to Moore's Law." *IEEE Transactions on Semiconductor Manufacturing* 24, no. 1: 1–4.

Smil, V. 2015. "Moore's Curse." *IEEE Spectrum,* April, 26.

GROWTH IN THE MODERN WORLD

Cunningham. C. 2020. "TV Screen Sizes over Time." VAVA, February 10. http://blog .vava.com/the-evolution-of-tv-screen-sizes-past-and-future-the-largest-4k-tv/.

European Commission. 2021. "Electricity Price Statistics." Eurostat. https://ec.europa .eu/eurostat/statistics-explained/index.php?title=Electricity_price_statistics#:~: text=The%20EU%20average%20price%20in,was%20%E2%82%AC0.2369%20 per%20kWh.

Feldman, D., et al. 2021. *U.S. Solar Photovoltaic System and Energy Storage Cost Benchmark: Q1 2020.* Technical Report NREL/TP-6A20-77324. US Department of Energy, National Renewable Energy Laboratory, January.

Kelly, B., et al. 2020. "Measuring Technological Innovation over the Long Run." NBER Working Paper 25266. Cambridge, MA: National Bureau of Economic Research.

Smil, V. 2005. *Creating the 20th Century: Technical Innovations of 1867–1914 and Their Lasting Impact.* New York: Oxford University Press.

Smil, V. 2006. *Transforming the 20th Century: Technical Innovations and Their Consequences.* New York: Oxford University Press.

Smil, V. 2016. *Still the Iron Age: Iron and Steel in the Modern World.* Amsterdam: Elsevier.

Smil, V. 2019. *Growth: From Microorganisms to Megacities*. Cambridge, MA: MIT Press.

UN Food and Agricultural Organization. 2021. *The State of Food Security and Nutrition in the World 2021*. Rome: FAO.

UN Food and Agriculture Organization. 2022. Crops and Livestock Products (database). Food and Agriculture Statistics, FAOSTAT.

World Bank. 2022. GDP Per Capita (Constant 2015 US$) (database). https://data.worldbank.org/indicator/NY.GDP.PCAP.KD.

Zu, C., and H. Li. 2011. "Thermodynamic Analysis on Energy Densities of Batteries." *Energy and Environmental Science* 4:2614–2625.

CANCER

American Cancer Society. 2021. *Cancer Treatment and Survivorship: Facts & Figures 2019–2021*. Atlanta: American Cancer Society, 2019. https://www.cancer.org/content/dam/cancer-org/research/cancer-facts-and-statistics/cancer-treatment-and-survivorship-facts-and-figures/cancer-treatment-and-survivorship-facts-and-figures-2019-2021.pdf.

Farrelly, C. 2021. "50 Years of the 'War on Cancer': Lessons for Public Health and Geroscience." *Geroscience* 43:1229–1235.

Memorial Sloan Kettering Cancer Center. 2021. "Mission Possible? Revisiting the 'War on Cancer' 50 Years Later." *MSK News,* Winter. https://www.mskcc.org/msk-news/winter-2020/mission-possible-revisiting-war-cancer-50-years-later.

National Cancer Institute. 2020. "Milestones in Cancer Research and Discovery." Washington, DC: National Institutes of Health.

National Cancer Institute. 2021. "National Cancer Act of 1971." Washington, DC: National Institutes of Health, February (last update). https://www.cancer.gov/about-nci/overview/history/national-cancer-act-1971.

Obama, B. 2009. Address to Joint Session of Congress. Remarks of President Barack Obama—Address to Joint Session of Congress. https://obamawhitehouse.archives.gov/the-press-office/remarks-president-barack-obama-address-joint-session-congress.

Rehemtulla, A. 2009. "The War on Cancer Rages On." *Neoplasia* 11:1252–1263.

Sporn, M. B. 1996. "The War on Cancer." *Lancet* 347:1377–1382.

von Eschenbach, A. C. 2003. "NCI Sets Goal of Eliminating Suffering and Death due to Cancer by 2015." *Journal of the National Medical Association* 95:637–639.

Weir, H. K., et al. 2015. "The Past, Present, and Future of Cancer Incidence in the United States: 1975 through 2020." *Cancer* 121:1827–1837.

White House. 2022. "Fact Sheet: President Biden Reignites Cancer Moonshot to End Cancer as We Know It." Press release, February 2.

DECARBONIZATION

Breakthrough Energy. 2021. "Breakthrough Energy Catalyst and Major Corporations Announce Partnership to Accelerate the Clean Energy Transition." https://www.breakthroughenergy.org/catalyst-announcement.

International Energy Agency. 2020. *Global EV Outlook 2020.* IEA, June. https://www.iea.org/reports/global-ev-outlook-2020.

International Energy Agency. 2021. "CO_2 Emissions: Global Energy Review 2021." IEA. https://www.iea.org/reports/global-energy-review-2021/co2-emissions.

Smil, V. 2017. *Energy Transitions: Global and National Perspectives.* Santa Barbara, CA: Praeger.

Smil, V. 2021. "SUVs Ascendant." *IEEE Spectrum,* September, 22–23.

Smil, V. 2021. "Electric Flight." *IEEE Spectrum,* November, 22–23.

UN Framework Convention on Climate Change, Glasgow Climate Pact. 2021. "Draft text on 1/CMA.3." November 13. https://unfccc.int/sites/default/files/resource/Overarching_decision_1-CMA-3_1.pdf.

INDEX

Note: Page numbers in italics indicate illustrations.